AF386721

Synthesis Lectures on Human-Centered Informatics

Series Editor

John M. Carroll, College of Information Sciences and Technology, Penn State University, University Park, USA

This series publishes short books on Human-Centered Informatics (HCI), at the intersection of the cultural, the social, the cognitive, and the aesthetic with computing and information technology. Lectures encompass a huge range of issues, theories, technologies, designs, tools, environments, and human experiences in knowledge, work, recreation, and leisure activity, teaching and learning, etc. The series publishes state-of-the-art syntheses, case studies, and tutorials in key areas. It shares the focus of leading international conferences in HCI.

Sergio Sayago

Older Adults and Digital Technologies

Background, Design, Configurations, and Futures

Springer

Sergio Sayago
Universitat de Lleida
Lleida, Spain

ISSN 1946-7680   ISSN 1946-7699 (electronic)
Synthesis Lectures on Human-Centered Informatics
ISBN 978-3-032-25555-6   ISBN 978-3-032-25556-3 (eBook)
https://doi.org/10.1007/978-3-032-25556-3

This Springer imprint is published by the registered company Springer Nature Switzerland AG
The registered company address is: Gewerbestrasse 11, 6330 Cham, Switzerland

If disposing of this product, please recycle the paper.

Preface

As a researcher deeply committed to exploring the human dimension of digital technologies, I am profoundly grateful to the participants who have made this work possible. I owe a special debt of gratitude to all the older adults who generously took part in my studies and shared their experiences with me. On more than one occasion, I recall hearing some of them saying that they were learning a great deal; it has been me who has learned tons with, and from, all of them. This book has been written with sincerity and dedication. It stands as a tribute to all of them.

I am also deeply thankful to those who have both believed in and supported me throughout my academic journey thus far. My heartfelt appreciation goes to Dr. Paula Forbes for her unwavering support. I consider myself fortunate to have Paula not only as a colleague but also as a friend. Thanks a lot, Paula.

I extend my gratitude to the institutions where I have had the privilege to develop my academic career: Universitat Pompeu Fabra, University of Dundee, Universidad Carlos III de Madrid, Universitat de Barcelona, and Universitat de Lleida. I am equally indebted to the educational, social, and community centres where I developed my fieldwork: Casal d'Avis 1er Maig, La Comunitat d'Aprenentatge Verneda-Sant Martí, The Dundee User Centre, and El Espacio Fundación "la Caixa" Madrid. This book represents the culmination of years of research conducted across these universities and centres.

Finally, I would like to thank Christine Kiilerich and her team at Springer for their editorial assistance and patience throughout this process. Writing this book has taken longer than anticipated. It has also required considerable effort. I am also grateful to the reviewers for their detailed, insightful, and constructive comments and suggestions.

Lleida, Spain   Sergio Sayago

Contents

Introduction

1

Abstract

This chapter presents the central objective of the book and the methodological approach taken to achieve it. It explains the motivations behind this synthesis and offers an overview of the book's structure, content, and key limitations. Additionally, it situates the work within a broader academic context and articulates its stance on adopting age-inclusive language when writing about, and working with, older adults.

Keywords

Older adults · Digital technologies · Synthesis · Reflective research · Age-inclusive language

1.1 Purpose and Approach

This book aims to provide a reflective synthesis of older-adult HCI research. It examines the diverse perspectives that shape this field and analyses how these viewpoints inform both current practices and future research perspectives. By critically reflecting on how studies are conceptualized, conducted, and reported, as well as the knowledge they produce, this book seeks to advance the field of older-adult HCI.

This synthesis is based on a critical review of more than 1000 publications, including conference proceedings and journal articles (see Table 1.1 for an overview), together with doctoral dissertations, technical reports and books. Most of these works have been published over the past three decades (starting in the 1990s). Given the inherently interdisciplinary nature of research on digital technologies and older adults, the review also draws on contributions from fields adjacent to HCI, as reflected in the references cited throughout the chapters.

© The Author(s), under exclusive license to Springer Nature Switzerland AG 2026

S. Sayago, *Older Adults and Digital Technologies*, Synthesis Lectures on Human-Centered Informatics, https://doi.org/10.1007/978-3-032-25556-3_1

Table 1.1 Overview of journals and conferences

Journals	Conferences
ACM Transactions on Accessible Computing	ACM-W4A (Web for Accessibility)
ACM Transactions on Computer–Human Interaction	ASSETS (ACM SIGACCESS Conference on Computers and
AI & Society	Accessibility)
Ageing and Mental Health	AutomotiveUI
Ageing and Society	British HCI
Ageing International	CHI
Ageing Research Reviews	Computers Helping People with Special Needs
American Behavioural Scientist	CSCW (Computer Supported Collaborative Work)
Annual Review of Sociology	DIS (Designing Interactive Systems)
Applied Ergonomics	DSAI (International Conference on Software Development and
BMC Geriatrics	Technologies for Enhancing Accessibility and Fighting
Behaviour & Information Technology	Info-exclusion)
Communications of the ACM	HCI International
Computers in Human Behaviour	IHRI (Human–Robot Interaction)
Current Opinion in Psychology	ITAP (International Conference on Human Aspects of it for the
Developmental Review	Aged Population)
Disability and Rehabilitation: Assistive Technology	MobileHCI
Educational Gerontology	NordiCHI
European Journal of Ageing	OzCHI
Foundations and Trends® in Human–Computer Interaction	Pervasive Health
Frontiers in Robotics and AI	PETRA (PErvasive Technologies Related to Assistive
Games and Culture	Environments)
Gerontechnology	
History and Philosophy of the Life Sciences	
IEEE Transactions on Human–Machine Systems	
Information, Communication & Society	
Interacting with Computers	
International Journal of Human–Computer Interactions	
International Journal of Human–Computer Studies	
International Journal of Social Robotics	
Journal of Aging Studies	
Journal of Engineering Design	
Journal of Sociology	
Journalism and Media	
New Media & Society	
Personal and Ubiquitous Computing	
PLOS ONE	
Psychology & Ageing	
Science	
Social Sciences	
Technological Forecasting and Social Change	
Telematics and Informatics	
The Design Journal	
The Gerontologist	
Universal Access Information Society	

This book also builds on the author's own research, encompassing 15 years of engagement with older adults in the context of HCI. This body of work includes ethnographic research, participant observation, and co-design studies that examine technology use and design with older individuals from diverse cultural and socio-economic backgrounds. Further insights are informed by the author's experience serving on editorial boards of recognized academic journals and programme committees of international conferences.

1.2 Context and Motivation

1.2.1 Why Focus on Older Adults?

This question requires a clear and well-founded justification. Within HCI, it has become common practice to cite the growing ageing population—now widely recognized as a societal reality—as a rationale for conducting research involving older adults. A similar trend is evident across other disciplines engaged with digital technologies and ageing, including Gerontology, Robotics, Media Studies, and Ethics.

Demographic changes undoubtedly provide strong reasons to investigate how older adults interact with and are affected by digital technologies. For example, in the domain of health care, which has been attracting growing research and social attention for a long time, (Schicktanz & Schweda, 2021) outline how existing ethical frameworks based on medical ethics need to be extended or reconsidered to capture the ethical issues posed by technological developments regarding care for older people. Within this context, it is also important to go beyond the traditional compensatory role that socially assistive robots (SAR) play in caring for older adults so that these robots can effectively recognise and respond to their values, ensuring a positive and sustained user experience while minimising abandonment (Simão et al., 2024). To fight against abandonment or non-acceptance of robots and other digital technologies, (Li et al., 2020) point out that stigma perception among older adults plays an important role; yet stigma perception is a poorly understood aspect of the *gerontechnology* design space.

With the advent of automated vehicles, ongoing research is looking into the future mobility of older people, which is imperative for maintaining well-being and quality of life in a growing ageing society. According to Li et al. (2021), despite the tendency in current research into older people and level 3 automated vehicles[1] to consider older adults as a homogenous group, this segment of the population is in reality highly diverse in terms of their performance, capabilities, needs and requirements when interacting with automated vehicles.

Recent advances in conversational AI and the ubiquity of related devices and applications have led to extensive research on designing and studying conversational systems with older adults. A recent systematic review (Huang et al., 2025) shows that while there are promising benefits these technologies can offer, concerns about privacy and trust predominate in older adults' perceptions of conversational AI.

These and other works, which are discussed throughout the book, highlight that population ageing presents the HCI community with a unique opportunity to examine how the field responds to a fundamentally new and unprecedented societal phenomenon. However, this synthesis argues that population ageing, and its frequently negative framing, should not be the sole justification for research focused on older adults. As self-evident as it may

---

[1] The driver can be completely disengaged from driving while, under some circumstances, being expected to take over the control occasionally.

seem, older people matter, just like any other user group. Independent of demographic arguments, older adults deserve digital technologies that are worthwhile to use, i.e. technologies that are non-stigmatising, accessible, usable, meaningful, and that promote both social and digital inclusion.

This objective is undoubtedly shared by many researchers in the HCI community (and others). Nevertheless, I wish to state it explicitly at the outset of this book, as much of the existing literature suggests that scholars (and reviewers) often feel compelled to justify the relevance of studying older adults' engagement with digital technologies, as though such inquiry was not inherently valuable.

1.2.2 Why a Reflective Synthesis?

Research contributions in HCI are not only empirical, theoretical or methodological (Wobbrock & Kientz, 2016). Studies, also called essays or arguments, which draw on many of the previous contribution types to make their case, and that compel reflection (Schon, 1983), discussion and debate, are another type of research contribution. Despite running the risk of providing an oversimplified list of exemplary studies, noteworthy research contributions of this type include the following seminal publications: "from human factors to human actors" by Bannon (1991), "implications for design" by Dourish (2006), "when the second wave meets the third one" by Bødker (2006), "HCI research as problem-solving" by Oulasvirta and Kasper (2016), and the analysis of how we involve people in design by Vines et al. (2013). More recent studies include (Lee et al., 2022), which examines what roles humans play in Human–Robot Interaction research, and (Agnew et al., 2024), which analyses proposals recommending the replacement of human participants in technology development and scientific research with LLMs. The first CHI workshop on meta-research (Oppenlaender et al., 2025) is also worth mentioning, as it argues that "it is time for the CHI community to pause and reflect on how things have progressed in the past decade and where we are headed".

As the older-adult HCI field grows, as evidenced by the extensive literature[2] on which this book builds, I consider that it is important to reflect on the underlying research to "look at something which was once constructed and may be reconstructed" (Schon, 1983, p. 296). The very act of defining, and writing about, a group as underrepresented or in need of assistance is an exercise of power that influences how that group is understood (Mylonopoulou et al., 2020) and can shape perceptions in ways that may not always be constructive (Sin et al., 2025). For example, patronizing discourses and pervasive negative stereotypes about older adults (Ng et al., 2015) hinder the development of a nuanced, fair, and accurate understanding of this population. There are studies that frame digital technology as a solution to most of the challenges faced by many older adults. Others, however, move beyond solutionist narratives to explore what digital technologies

---

[2] Which is dispersed across a wide range of publication venues.

might afford them. This raises an important question: which approach or combination of approaches should be adopted, and why? Much existing research treats ageing as a static condition; however, ageing is a dynamic and ongoing process of gains and losses. Technology design approaches vary significantly, from deficit-oriented perspectives to those based on principles of positive design—each carrying the potential, often unintentionally, to reinforce age-related stereotypes. Older-adult HCI research frequently mirrors the latest technological trends. Whereas exploring different technologies allows us to get a better understanding of issues that matter to older adults, the field seems to start from scratch with every new technology. How do we use the knowledge accumulated over decades of research to design emerging digital technologies? The field of HCI is increasingly focused on designing technologies to address sustainability. How can older-adult HCI research contribute to post-growth goals (Sharma et al., 2024)? Such concerns highlight the need for ongoing critical analysis and reflection.

Another motivation for this synthesis lies in the centrality of ageing within our increasingly hybrid everyday lives, where the boundaries between technology and humans are increasingly fuzzy (Frauenberger, 2020), and interactions occur not only online and offline but also in relation to non-human entities such as sensors, pets, and algorithms. None of us ages alone. We all age with other humans, with other living creatures, and with things (Lupton, 2024). As stated by Pradhan et al. (2025), recent work in critical gerontology (Andrews & Duff, 2019) is beginning to question whether a sole humanistic focus can provide a holistic perspective on technologies for ageing. More-than-human perspectives are gaining ground in HCI research (Eriksson et al., 2024). Yet, more-than-human-ageing perspectives are yet to be examined in older-adult HCI research.

1.3 Overview of the Book

This book engages with issues that are relevant to the study of digital technologies and older adults—regardless of the specific domain of inquiry.

Age, Ageing, and Older Adults

This synthesis argues that while chronological age—the number of years a person has lived—is often treated as a key variable in this field, it is far from sufficient to explain the rich and varied relationships older individuals have with digital technologies. Age is more than a number; it encompasses a dynamic process of change across multiple dimensions. As people grow older, they accumulate experiences and undergo transformations; some within their control, others shaped by broader societal forces. Yet, older-adult HCI research frequently conceptualizes ageing as a static state or a fixed point in time.

This book also critiques the long-standing tendency to treat older adults as a homogenous group, a framing that overlooks the diversity within this population. At the same time, it cautions against overemphasizing heterogeneity, which can obscure shared experiences and common needs that are essential to understanding older adults as a distinct user group. Striking a balance between recognizing diversity and identifying meaningful commonalities is crucial for advancing inclusive technologies and effective design practices.

Ways of Designing for Older Adults
Do older adults truly need simpler digital technologies, or should we instead focus on designing complexity more thoughtfully? Are their "special" needs, such as independence, competence, and inclusion, fundamentally different from those of other user groups? And if technology eventually succeeds in addressing the challenges associated with ageing, what remains of the experience of ageing and later life?

This book invites the older-adult HCI community to reflect critically on how technologies for older adults are designed and on the consequences of our work. The literature identifies three prominent design approaches: deficit-driven design, positive design, and design based on technology generation and familiarity. These approaches differ significantly in their objectives and in how they conceptualize older adults, and their implications deserve careful attention.

For instance, do design strategies that compensate for declining capabilities or rely on generational familiarity inadvertently reinforce stereotypes? Conversely, do approaches that emphasize empowerment, pleasure, and curiosity, which are often associated with the so-called "third age", risk marginalizing those in the "fourth age", where frailty and dependency may be more pronounced? How can we imagine older-centred design towards advancing sustainable futures?

These questions are central to the broader inquiry this book undertakes. They challenge us to move beyond simplistic or binary framings of ageing and to consider how design can more meaningfully engage with the complexity of later life.

Methodological Configurations
HCI is a rapidly evolving field requiring continual reflection on and adaption of our methodological approaches. As the user base grows ever more diverse, we are not just simply using digital technologies; we are living with them. Consequently, research is now being conducted in increasingly sensitive contexts and addresses more sensitive topics.

The field of older-adult HCI is not immune to methodological difficulties. In fact, there is growing evidence highlighting the challenges of conducting research with older participants. Yet, this book argues that scholarly attention to these issues remains limited and fragmented. Greater methodological reflection and transparency, particularly in the methods and discussion sections of publications, are needed to better understand when, how, and why specific methodological configurations are appropriate. This kind of reflexivity, already established within HCI, could be further developed in research with older adults.

The increasing diversity among older research participants also calls into question the one-size-fits-all approach that dominates much of the literature on research methods. While standardized methodologies can help guide researchers and support comparison and replicability, they may also obscure important differences and limit responsiveness to participants' varied needs and contexts. This tension between methodological consistency and participant diversity deserves closer scrutiny and more nuanced discussion within the older-adult HCI community.

Knowledge Production in HCI and Ageing

This synthesis highlights a recurring pattern in HCI research on older adults: the tendency to mirror the latest technological developments. When a new technology emerges, it often prompts a wave of studies exploring its relevance to older users. A clear example are conversational AI technologies and devices. Meanwhile, earlier technologies—and the unresolved challenges associated with them—tend to fade from view. This cycle risks leaving important issues unaddressed and contributes to a fragmented understanding of older adults' long-term engagement with digital technologies.

Another consequence of this mode of knowledge production is the limited attention paid to enduring concerns that matter deeply to many older people, such as privacy and trust. These topics are frequently mentioned in publications but rarely examined in depth. This book argues for a more sustained and critical engagement with such issues, without overlooking the relationship between older adults and emerging technologies. The overall objective is to find a balance between research driven by technological developments, which is undoubtedly valuable in today's fast-paced digital environment, and research that addresses issues that play an important role in technology use by older adults, and that demands further reflection and comprehensive understanding to make a long-lasting impact.

The field also exhibits a tension in how older adults are conceptualized. On the one hand, they are viewed through the lens of age-related changes in functional capabilities; on the other, they are seen as active, socially embedded individuals and discerning users of technology. Each perspective shapes the research questions posed, the methods employed, and the nature of the relationship between older adults and digital technologies. While one perspective may be implicitly privileged over the other, neither is inherently superior or incorrect. Rather, they reflect different epistemological orientations that could be made more explicit and critically examined in future research.

Digital Technologies and the Everyday Lives of Older Adults

Each new digital technology is often accompanied by enthusiasm about its potential to enhance the everyday lives of older adults. However, despite decades of research, evidence regarding the impact of digital technologies on their quality of life, subjective well-being, and independence remains inconclusive and mixed.

This synthesis argues for a clearer distinction between evidence-based claims and utopian or overly optimistic narratives about the transformative potential of digital technologies. While optimism can drive innovation, it could be tempered by rigorous and ongoing research that critically examines the actual effects of technology on older adults' lives.

To this end, continued and improved research is essential—not only to assess impact but also to ensure that technologies are designed and implemented in non-ageist ways that support older adults' needs, preferences, and lived experiences.

Future Perspectives

As we look ahead, the design and study of digital technologies for older adults may benefit from expanding beyond a purely human-centred lens to consider the broader ecosystem of human and non-human actants. Repositioning older adults within this hybrid network, where interactions involve not only people but also things and other living creatures, raises important questions: Where might such a shift take us? How will it shape future HCI research concerning ageing?

If we accept that ageing is a continuous process of change, and that we are likely to live longer lives, then HCI must evolve to reflect this demographic reality. This may require a cultural shift in how we conceptualize ageing and design technologies that support individuals across ten or more decades of life. Strategic thinking is needed to determine how best to leverage HCI knowledge to enable longer, richer, and more hybrid lives.

Moreover, the next wave(s) of older adults—shaped by different technological, cultural, and social experiences—will likely redefine what it means to age and engage with technology. This evolution will challenge existing assumptions and open new avenues for research, design, and policy. Understanding how these shifts will transform the nature and scope of older-adult HCI is a key concern for the future.

1.4   Contribution

To situate the contribution this book aims to make to research on digital technologies and older adults within a broader scholarly context, selected prior, and related works are reviewed in Sect. 1.4.1.

1.4.1   Selected Previous Works

This synthesis is not the first research book to address the topic of digital technologies and older adults. Several notable works have laid important foundations in the field. For instance, Fisk et al., (2004, 2009) produced a seminal publication grounded in extensive empirical research in human factors. Johnson and Finn (2017) offer a clear, provocative, and accessible exploration of user interface design for an ageing population. McLaughlin and Pak (2020) focus on the fundamental changes that tend to accompany ageing (physical, social, and cognitive) and translate the latest research in cognitive ageing into recommendations and principles for designing displays for older adults. Prendergast and Garattini (2015) approach the topic from a life course perspective, emphasizing how "technologies, sensitively designed with social, personal and institutional practices in mind, can help contribute towards building positive experiences in the later life course" (p. 1). Sixsmith and Gutman (2013) focus on the concept of active ageing, a central theme in both public discourse and research concerning older adults and digital technologies and examine the extent to which technology can support it. Finally, Neves and Vetere (2019) present a highly interdisciplinary volume that brings together scholars from the social and computer sciences to explore theoretical, methodological, ethical, and empirical approaches to designing technologies that respond to the needs, desires, and aspirations of older adults.

Previous publications have also taken stock of HCI research involving older adults in specific areas and topics of interest. For example, Marzi et al. (2023) explore the motivations and needs of older adults regarding their participation in HCI research, finding that overall willingness to participate is influenced by factors like age-related illnesses, perceived lack of skills, and the desire to avoid commitments during retirement. Cozza et al. (2017) examine how ubicomp research published in the *Journal of Personal and Ubiquitous Computing* (1997–2014) has portrayed the relationship between technologies and older users. Their analysis reveals a dominant emphasis on the functional paradigm, which assumes that technology enables older people to live independently, while often defining older users primarily in terms of frailty and dependency. Burema (2022) critiques representations of older adults in the field of Human–Robot Interaction, arguing that they are often depicted as "frail by default, independent by effort, silent, and technologically illiterate, burdensome and problematic for society (...) The older body is seen as fixable with social robots". Following up on robots, Mannheim et al. (2022) argue that ageist and interventionist perspectives on the imagined reality of being older seem to dictate the primary use case of social assistive robots for care and alleviation of loneliness. Knowles and Hanson (2018a) advocate for greater attention to the non-use of digital technologies by older adults, noting that "much of what underlies the resistance is inherent in aspects of being older that are unlikely to change as new generations reach retirement age". They suggest that non-use can be instructive for identifying problematic consequences of digital technologies and for informing redesign. In a related publication, Knowles and Hanson (2018b) explore the role of distrust in technology adoption, proposing that distrust should be understood in relation to values and normative expectations.

Other works have examined HCI research with older people from a broader perspective. These works have become—at least, in my opinion—seminal studies, giving shape to older-adult HCI (Moffatt, 2013) over the last years. Vines et al. (2015) delineate an age-old problem in HCI, wherein ageing is typically framed as a problem that can be managed by technology. Newell and Gregor (2002) argue that there is little evidence that older people are particularly technophobic. They also point out that there is a clear social and demographic imperative to design systems that are appropriate for use by older people and for those with disabilities. Rogers and Marsden (2013) challenge the tendency to develop technological solutions for older people by providing for a lack of something and encouraging us to promote empowerment through technology. Moffatt (2013) highlights the need "to start thinking about ways to strengthen our community identity, not just for the benefit of bringing like-minded researchers together but also for the purpose of clearly defining and communicating the breadth and depth of the field to the outside world". Petrie (2023) argues that research on digital technologies for older people is largely failing to address two key issues, the tendency to treat all older people as a homogenous group, and the lack of understanding and stereotypical views about older people by young researchers and developers. Light et al. (2016) remind us that people do not make a choice between dependence and independence. They experience a gradual re-evaluation of their roles. In another publication also led by Light (2015), the authors argue for a departure from the deficit model of old age to an understanding that reveals older people's agency in the ageing process and the work they do to manage their capacity to age well. More recently, Lazar et al. (2025) examine a critical turn on research with older adults in HCI that emerged in the 2010s by focusing on technology use, intersectionality and care.

1.4.2   Key Contribution

This book provides an interpretative synthesis of extensive interdisciplinary research on digital technologies and older adults. It strengthens existing themes in the literature by presenting new evidence, while also highlighting critical issues, tensions, and questions that merit further exploration. In doing so, it aims to support HCI scholars—across all levels of academic seniority—in recognizing the diverse perspectives that coexist within the field, understanding how these viewpoints shape current research, and reflecting on how they might inform future inquiries into ageing and digital technology.

1.5   Age-Inclusive Language

"Words or phrases can suggest bias or reflect negative, disparaging, or patronizing attitudes towards individuals or groups of individuals. These words and phrases can influence our impressions, attitudes, and even our actions. Choosing language that represents the preference of the groups to which it refers can convey respect and integrity" (Hanson et al., 2015).

Negative stereotypes about older adults are present in writing and everyday speech.[3] Our choice of language contributes to a pervasive cultural narrative that ageing is something to be feared or ashamed of. As HCI scholars, I consider that we are well placed to lead by example when writing about older people in a person-centred, fair and accurate way.

This book aims to contribute to the Reframing Ageing initiative[4] by using an age-inclusive language.

Guidelines[5] from the American Psychology Association (APA)[6] and the American Medical Association,[7] amongst others, advise against using "elderly", "senior", or "the aged". These terms connote a stereotype and suggest that members of the group are not part of society but rather a group apart. Terms such as "older persons", "older people", "older adults", "older individuals", "persons 65 years and older", and "the older population", are preferred. The word adult affirms agency and personhood, as does person-first language (Bowman & Lim, 2021). The rise of named cohorts like "Boomers" or "Granfluencers" (Ng & Indran, 2023) reinforces ageism, because of the identification with a cultural group (Morganroth, 2004). Generational descriptors such as "baby boomers", "Gen X", "millennials", "centennials", "Gen Z" and so on should be used only when discussing studies related to the topic of generations.

1.6   Limitations

This synthesis does not constitute a systematic review, bibliometric analysis, or meta-research study. It does not aspire to the levels of transparency, replicability, or comprehensiveness typically expected of these types of research contributions. Although this may be perceived as a limitation, the primary aim is to offer a reflexive and interpretive synthesis of the literature, grounded in over a decade of experience conducting HCI research with older adults. All publications included in this book were read in full and examined as thoroughly as possible.

This synthesis may not provide definitive answers to the complex issues it raises. Addressing these challenges requires a collective effort that extends beyond the scope of a single publication. Some of the suggestions and future directions proposed here may be tentative or speculative. They are intended to provoke thought and inspire further

---

[3] https://publichealth.wustl.edu/age-inclusive-language-are-you-using-it-in-your-writing-and-eve ryday-speech/ (Access: 16/09/2025).

[4] https://www.reframingaging.org/ (Access: 16/09/2025).

[5] https://publichealth.wustl.edu/age-inclusive-language-are-you-using-it-in-your-writing-and-eve ryday-speech/ (Access: 16/09/2025).

[6] https://apastyle.apa.org/style-grammar-guidelines/bias-free-language/age (Access: 16/0972025).

[7] https://academic.oup.com/amamanualofstyle/book/27941/chapter/207567296?login=true#529 876675 (Access: 16/09/2025).

research. Whether one agrees or disagrees with the ideas put forward, the opportunity to look ahead is essential for the continued growth of older-adult HCI. New perspectives inevitably prompt change, and it is likely that this book will inspire alternative directions that will unfold in the years to come.

This book does not present design guidelines, nor does it engage in a focused discussion on health, technology, and older adults—two areas that have received substantial attention in existing research. While omitting these topics may be viewed as a limitation, the intention is to offer a different perspective on older-adult HCI, one that is more holistic in nature. Rather than concentrating on specific domains or prescriptive design advice, this synthesis aims to foster broader reflection on key aspects that characterise our work in this field.

This book has been written primarily from a European perspective. While the research literature it draws upon is international, it reflects a strong European and North American orientation. Given that ageing is a universal experience—and one deeply shaped by cultural contexts—certain views on ageing and digital technology, as well as research conducted in other regions, may be underrepresented or even overlooked. Acknowledging this limitation, future work could help fill this gap by incorporating more diverse cultural perspectives and expanding the geographical scope of older-adult HCI research.

1.7   Organisation

This book is organized into five chapters. Chapter 2 provides an interdisciplinary overview of key concepts in the research literature and reflects on their relevance within the context of older adults and digital technologies. Chapter 3 focuses on older-adult-centred design, exploring different ways of understanding older adults and technologies, and approaches to designing for this population. Chapter 4 builds on the previous chapter by examining methodological configurations in research involving older adults. Chapter 5 concludes the book with an invitation to reflect on possible future research directions that the field of older-adult HCI might pursue. Chapter 6 presents the complete list of references of this book.

References

Agnew, W., Bergman, A. S., Chien, J., Díaz, M., El-Sayed, S., Pittman, J., Mohamed, S., & McKee, K. R. (2024). The illusion of artificial inclusion. In *Proceedings of the CHI conference on human factors in computing systems* (pp. 1–12). https://doi.org/10.1145/3613904.3642703

Andrews, G., & Duff, C. (2019). Understanding the vital emergence and expression of aging: How matter comes to matter in gerontology's posthumanist turn. *Journal of Aging Studies, 49*, 46–55. https://doi.org/10.1016/j.jaging.2019.04.002

Bannon, L. J. (1991). From human factors to human actors the role of psychology and human-computer interaction studies in systems design. In J. Greenbaum & M. Kyng (Eds.), *Design at work: Cooperative design of computer systems* (pp. 25–44). Lawrence Erlbaum Associates.

Bødker, S. (2006). When second wave HCI meets third wave challenges. In: *NordiCHI* (pp 14–18). Oslo, Norway.

Bowman, C., & Lim, W. M. (2021). How to avoid ageist language in aging research? An overview and guidelines. *Activities, Adaptation & Aging, 45*(4), 269–275. https://doi.org/10.1080/019 24788.2021.1992712

Burema, D. (2022). A critical analysis of the representations of older adults in the field of human–robot interaction. *AI & SOCIETY, 37*(2), 455–465. https://doi.org/10.1007/s00146-021-01205-0

Cozza, M., De Angeli, A., & Tonolli, L. (2017). Ubiquitous technologies for older people. *Personal and Ubiquitous Computing, 21*(3), 607–619. https://doi.org/10.1007/s00779-017-1003-7

Dourish, P. (2006). Implications for design. In *Proceedings of the SIGCHI conference on human factors in computing systems (CHI '06)* (pp. 541–550). Association for Computing Machinery. https://doi.org/10.1145/1124772.1124855

Eriksson, E., Yoo, D., Bekker, T., & Nilsson, E. M. (2024). More-than-human perspectives in human-computer interaction research: A scoping review. In *Nordic conference on human-computer interaction* (pp. 1–18). https://doi.org/10.1145/3679318.3685408

Fisk, A., Rogers, W., Charness, N., Czaja, S., & Sharit, J. (2004). Designig for older adults. In *Principles and creative human factors approaches* (1st ed.). CRS Press.

Fisk, A., Rogers, W., Charness, N., Czaja, S., & Sharit, J. (2009). Designig for older adults. In *Principles and creative human factors approaches* (2nd ed.). CRS Press.

Frauenberger, C. (2020). Entanglement HCI the next wave? *ACM Transactions on Computer-Human Interaction, 27*(1), 1–27. https://doi.org/10.1145/3364998

Hanson, V. L., Cavender, A., & Trewin, S. (2015). Writing about accessibility. *Interactions, 22*(6), 62–65. https://doi.org/10.1145/2828432

Huang, Y., Zhou, Q., & Piper, A. M. (2025). Designing conversational AI for aging: A systematic review of older adults' perceptions and needs. In *Proceedings of the 2025 CHI conference on human factors in computing systems* (pp. 1–20). https://doi.org/10.1145/3706598.3713578

Johnson, J., & Finn, K. (2017). *Designing user interfaces for an aging population: Towards universal design*. Elsevier Science.

Knowles, B., & Hanson, V. L. (2018a). Older adults' deployment of 'distrust.' *ACM Transactions on Computer-Human Interaction, 25*(4), 1–25. https://doi.org/10.1145/3196490

Knowles, B., & Hanson, V. L. (2018b). The wisdom of older technology (non)users. *Communications of the ACM, 61*(3), 72–77. https://doi.org/10.1145/3179995

Lazar, A., Brewer, R. N., & Knowles, B. (2025). HCI and older adults: The critical turn and what comes next. *Foundations and Trends® in Human-Computer Interaction, 19*(2), 112–212. https://doi.org/10.1561/1100000094

Lee, H. R., Cheon, E., Lim, C., & Fischer, K. (2022). Configuring humans: What roles humans play in HRI research. In *2022 17th ACM/IEEE international conference on Human-Robot Interaction (HRI)* (pp. 478–492). https://doi.org/10.1109/HRI53351.2022.9889496

Li, C., Lee, C.-F., & Xu, S. (2020). Stigma threat in design for older adults. *International Journal of Design, 14*(1).

Li, S., Blythe, P., Zhang, Y., Edwards, S., Xing, J., Guo, W., Ji, Y., Goodman, P., & Namdeo, A. (2021). Should older people be considered a homogeneous group when interacting with level 3 automated vehicles? *Transportation Research Part f: Traffic Psychology and Behaviour, 78*, 446–465. https://doi.org/10.1016/j.trf.2021.03.004

Light, A., Leong, T. W., & Robertson, T. (2015). Ageing Well with CSCW. In N. Boulus-Rødje, G. Ellingsen, T. Bratteteig, M. Aanestad, & P. Bjørn (Eds.), *ECSCW 2015: Proceedings of the*

14th European conference on computer supported cooperative work, 19–23 September 2015, Oslo, Norway (pp. 295–304). Springer International Publishing. https://doi.org/10.1007/978-3-319-20499-4_16

Light, A., Pedell, S., Robertson, T., Waycott, J., Bell, J., Durick, J., & Leong, T. W. (2016). What's special about aging. *Interactions, 23*(2), 66–69. https://doi.org/10.1145/2886011

Lupton, D. (2024). Towards a gerontology of everything: A more-than-human perspective. *Journal of Aging Studies, 71*, Article 101278. https://doi.org/10.1016/j.jaging.2024.101278

Mannheim, I., Wouters, E. J. M., Köttl, H., van Boekel, L., Brankaert, R., & van Zaalen, Y. (2022). Ageism in the discourse and practice of designing digital technology for older persons: A scoping review. *The Gerontologist,* gnac144. https://doi.org/10.1093/geront/gnac144

Marzi, K., Klapperich, H., & Huldtgren, A. (2023). Insights on older adults' willingness, motives and experiences regarding participation in HCI research. *Mensch und Computer, 2023*, 432–436. https://doi.org/10.1145/3603555.3608540

McLaughlin, A., & Pak, R. (2020). *Designing displays for older adults (Second edition).* CRC Press.

Moffatt, K. (2013). Older-adult HCI: Why should we care? *Interactions, 20*(4), 72–75. https://doi.org/10.1145/2486227.2486242

Morganroth, M. (2004). *Aged by culture.* The University of Chicago Press.

Mylonopoulou, V., Weilenmann, A., Torgersson, O., & Jungselius, B. (2020). Searching for empathy: A Swedish study on designing for seniors. In *Proceedings of the 11th Nordic conference on human-computer interaction: Shaping experiences, shaping society* (pp. 1–10). https://doi.org/10.1145/3419249.3420125

Neves, B. B., & Vetere, F. (Eds). (2019). *Ageing and digital technology: Designing and evaluating emerging technologies for older adults.* Springer Singapore. https://doi.org/10.1007/978-981-13-3693-5

Newell, A. F., & Gregor, P. (2002). Design for older and disabled people—Where do we go from here? *Universal Access in the Information Society, 2*(1), 3–7.

Ng, R., & Indran, N. (2023). Granfluencers on TikTok: Factors linked to positive self-portrayals of older adults on social media. *PLoS ONE, 18*(2), Article e0280281. https://doi.org/10.1371/journal.pone.0280281

Ng, R., Allore, H. G., Trentalange, M., Monin, J. K., & Levy, B. R. (2015). Increasing negativity of age stereotypes across 200 years: Evidence from a database of 400 Million words. *PLoS ONE, 10*(2), Article e0117086. https://doi.org/10.1371/journal.pone.0117086

Oppenlaender, J., Malacria, S., Fang, X., Van Berkel, N., Chevalier, F., Yatani, K., & Hosio, S. (2025). Meta-HCI: First workshop on meta-research in HCI. In *Proceedings of the extended abstracts of the CHI conference on human factors in computing systems* (pp. 1–8). https://doi.org/10.1145/3706599.3706723

Oulasvirta, A., & Hornbæk, K. (2016). HCI research as problem-solving. In *Proceedings of the 2016 CHI Conference on Human Factors in Computing Systems (CHI '16)* (pp. 4956–4967). Association for Computing Machinery. https://doi.org/10.1145/2858036.2858283

Petrie, H. (2023). Talking 'bout my generation... or not?: The digital technology life experiences of older people. In *Extended abstracts of the 2023 CHI conference on human factors in computing systems* (pp. 1–9). https://doi.org/10.1145/3544549.3582742

Pradhan, A., Chopra, S., Upadhyay, P., Brewer, R., & Lazar, A. (2025). Towards accounting for non-human agency in technology design for aging. In *Proceedings of the 18th ACM international conference on PErvasive technologies related to assistive environments* (pp. 273–276). https://doi.org/10.1145/3733155.3733205

Prendergast, D., & Garattini, C. (2015). *Aging and the digital life course (First edition).* Berghahn books.

Rogers, Y., & Marsden, G. (2013). Does he take sugar?: Moving beyond the rhetoric of compassion. *Interactions, 20*(4), 48–57. https://doi.org/10.1145/2486227.2486238

Schicktanz, S., & Schweda, M. (2021). Aging 4.0? Rethinking the ethical framing of technology-assisted eldercare. *History and Philosophy of the Life Sciences, 43*(3), 93. https://doi.org/10.1007/s40656-021-00447-x

Schön, D. (1983). *The reflective practitioner. How professionals thinks in action.* Basic Books.

Sharma, V., Kumar, N., & Nardi, B. (2024). Post-growth human-computer interaction. *ACM Transactions on Computer-Human Interaction, 31*(1), 1–37. https://doi.org/10.1145/3624981

Simão, H., Bernardino, A., Guerreiro, T., & Forlizzi, J. (2024). Beyond assistance: Robots aligned with older adults' values. In *Companion of the 2024 ACM/IEEE international conference on human-robot interaction* (pp. 145–147). https://doi.org/10.1145/3610978.3638357

Sin, J., Conte, S., Munteanu, C., Maitland, J., Nurain, N., Sarcar, S., et al. (2025). Beyond deficit-based design: Re-imagining co-creation approaches with equity-denied groups. In *Proceedings of the extended abstracts of the CHI conference on human factors in computing systems* (pp. 1–5). https://doi.org/10.1145/3706599.3716301

Sixsmith, A., & Gutman, G. (Eds). (2013). *Technologies for active aging.* Springer US. https://doi.org/10.1007/978-1-4419-8348-0

Vines, J., Clarke, R., Wright, P., McCarthy, J., & Olivier, P. (2013). Configuring participation: On how we involve people in design. In *Proceedings of the SIGCHI conference on human factors in computing systems* (pp. 429–438). https://doi.org/10.1145/2470654.2470716

Vines, J., Pritchard, G., Wright, P., Olivier, P., & Brittain, K. (2015). An age-old problem: Examining the discourses of ageing in HCI and strategies for future research. *ACM Transactions on Computer-Human Interaction, 22*(1), 1–27. https://doi.org/10.1145/2696867

Wobbrock, J. O., & Kientz, J. A. (2016). Research contributions in human-computer interaction. *Interactions, 23*(3), 38–44. https://doi.org/10.1145/2907069

Background: Age, Ageing, and Older Adults

Abstract

Terms such as *age*, *ageing*, and *older adults* are central to the body of research explored in this book. This chapter offers an interdisciplinary overview of these concepts and examines how they are framed within older-adult HCI research. Its aim is to lay the groundwork for the discussions on older adults and digital technologies that follow in the subsequent chapters of this book. Additionally, it aims to serve as an accessible entry point for HCI scholars interested in the study of ageing, older adults, and digital technologies. To achieve this, the chapter draws on research contributions from the last twenty years or so.

Keywords

Chronological age · Older age · Ageing · Older adults

2.1　Age

This section focuses on chronological age, i.e. the number of years since a person's birth. Despite playing a pivotal role in our research, a person's age is much more than chronology, and therefore this type of age is not useful enough to understand older people's experiences of technology use. This section also explores the concept of older age. Contrary to popular belief, older age is neither a uniform nor a stereotypical experience. This section underscores the potential of HCI research to challenge prevailing and often negative perceptions of it.

S. Sayago, *Older Adults and Digital Technologies*, Synthesis Lectures on Human-Centered Informatics, https://doi.org/10.1007/978-3-032-25556-3_2

2.1.1 Chronological Age

Chronological age plays a central role in how individuals perceive and categorize one another (Johfre & Saperstein, 2023). Alongside race and gender, age is one of the primary social markers used to quickly differentiate and group people (Rothermund & De Paula Couto, 2024). In most developed countries, the age of 65 is commonly used as the threshold for defining older adults. However, this cut-off is largely arbitrary (Czaja & Weingast, 2020).

In HCI research, chronological age is often treated as a convenient and foundational concept when studying older adults. Yet, and despite its intuitive appeal, chronological age should not be considered the sole determinant in older-adult HCI. There is much more to ageing than the number of years lived (Rothermund & De Paula Couto, 2024). For example, as people age, they shape their world in ways that maximize their well-being, and they do so within the confines and definitions of their respective cultures[1] (Fung, 2013).

As discussed by Schaie et al. (2011, 2016), and widely acknowledged in the academic literature, it is unlikely that any single variable can fully and exclusively account for the behaviour of older adults. This insight also applies to the relationship between older people and digital technologies (see Fig. 2.1). To illustrate this point, consider the growing body of research on digital technology acceptance and adoption among older adults. This is a highly complex area in which age is just one of many influencing factors. Others include social relationships, economic costs, daily routines and practices, and concerns related to trust, privacy, independence, and security, e.g. (Lujikx et al., 2015; Lee et al., 2019; Leonardi et al., 2009; Renaud & van Biljon, 2008; Wang & Sun, 2016; Macedo, 2017; Guner & Acarturk, 2020).

If age is not the sole factor that matters, older-adult HCI research risks losing what makes it distinct. However, this challenge can also be reframed as an opportunity to embrace a more nuanced and expansive understanding of age. As Johfre and Saperstein (2023) argue, age "is constructed through interwoven processes at the individual, interactional, organizational, and cultural levels, and an individual's age is far more than chronology". In other words, as a social construct, we can talk about chronological age but also about psychological (e.g. "I feel old") or social (e.g. retirement) age (Camacho-Markina & Santos-Diez, 2025; Chulián Horrillo et al., 2024).

Chronological age may be useful in quantitative HCI research, as it facilitates the grouping of participants. In contrast, its relevance is more contested in qualitative studies, which often focus on the lived digital experiences of older adults (Ashcroft & Knowles, 2025). For example, if we set out to design a new digital application for intergenerational communication, which description offers more insight from a user experience

---

[1] Culture is not synonymous of nationality. For something to be considered cultural, it must be learned, shared, and emerge in interaction. We can also talk about cultural groups beyond national ones, such as the Deaf people (Sayago, 2023a, 2023b).

Fig. 2.1 Chronological age does not help us to account for the behaviour of these two older research participants. They are above 65. What are they doing? Why are they fixing the printer? Will they succeed? It is important to know their past experiences and current aspirations to answer these and related questions—yes, they did fix the all-in-one-printer. It was a problem with the black ink cartridge. Photo by the author

design perspective: that our older research participants are grandparents (of any age) living in Scotland or Spain, with prior experience using computers, and with grandchildren living far away, or that they are aged 65 and above?

There is broad consensus that being above or below a certain chronological age reveals little about a person's daily activities, interests, or capabilities. Among older adults, we find a wide range of lifestyles, from highly socially active individuals to those who are homebound. This understanding of age, which goes beyond mere chronology, aligns with historical conceptions of older age, which have often been defined more by physical condition than by numerical age (Thane, 2005).

The life course of an older adult, along with life stages or transitions such as retirement, grandparenthood, and periods of autonomy or dependency, has been proposed as a more meaningful alternative[2] to chronological age for understanding later life, e.g. (Poortman and Van Tilburg, 2005; Cabrera & Malanowski, 2009; Heckhausen et al., 2010; Durrant et al., 2017; Barros et al., 2021; Barrett, 2022; Chulián Horrillo et al., 2024). These alternative frameworks for exploring later life remain relatively underdeveloped in HCI research involving older adults, e.g. (Barros et al., 2021; Caldeira et al., 2022; Jones et al., 2023; Lindley and Wallace, 2015; Smith, 2025; Willatt et al., 2024), where chronological age continues to dominate.

---

[2] However, this classification is not without its limitations. As Cruikshank (2009) points out, dependency in older age may not differ significantly from dependency experienced at other stages of life.

2.1.2 Older Age and Stereotypes

As stated in the seminal book *"The long history of old age"* (Thane, 2005), older age is commonly understood to span from around 50 to well beyond 100 years—a range unmatched by any other phase of life. Yet, despite its inevitability as part of the human lifecycle, older age is often overlooked or rendered invisible in societal discourse. A striking example is the lyric "Hope I die before I get old" from the song *My Generation*, which reflects a broader cultural tendency to devalue older age. This sentiment is reinforced by the prevailing cult of youth in most contemporary societies (Stark, 2020).

Older age is a natural stage of the human lifecycle. Yet, it has been surrounded by persistent myths and stereotypes for a long time (Victor, 1994). Contemporary narratives about ageing often reflect generalized and reductive views of older adults, including assumptions of: (1) poor health; (2) diminished mental acuity; (3) sadness, depression, and loneliness; (4) lack of vitality and sexual interest; and (5) an inability to learn or adapt (Thornton, 2002). These stereotypes are typically acquired early in life and, over time, become internalized. They shape individuals' perceptions of their own ageing and can act as self-fulfilling prophecies, influencing the actual development and behaviour of older adults (Rothermund & De Paula Couto, 2024).

Negative stereotypes about older adults are widespread. For example, the media do not sufficiently make older adults visible. It often presents negative narratives about old age and uses stigmatizing terms to refer to older adults. Diversity among older people is occasionally reflected (Bossio et al., 2023; Camacho-Markina & Santos-Diez, 2025). In the context of digital technologies, HCI literature has identified several recurring assumptions, for example, that older people are uninterested in technology or incapable of learning to use it (Durick et al., 2013). However, such age-based representations are often disconnected from the lived experiences of many older adults (see Fig. 2.1) (Ghorayeb et al., 2021; Li & Waycott, 2023; Sayago, 2019; Von Humboldt et al., 2024).

HCI research is well positioned to challenge these stereotypes and disrupt the status quo (Von Humboldt et al., 2024). Indeed, a growing number of HCI studies actively contests negative portrayals of older adults and their relationship with digital technologies, e.g. (Brewer & Piper, 2016; Kannabiran et al., 2020; Rosales & Fernández-Ardèvol, 2016; Tang et al., 2023; Waycott et al., 2013; Yang & Moffatt, 2024; Zhang et al., 2020; Zhao et al., 2024). It might be worth noting that older adults can be resilient and adapt to challenging life circumstances, and this was the case with new technologies during the pandemic. Despite some initial onboarding difficulties, most learned and adapted to additional channels for virtual communication (Wang et al., 2024). Moreover, some tech savvy older adults act as guardians of their communities for privacy protection (Nicholson et al., 2021), contradicting prior deficit-based narratives in characterizing older adults (Zou et al., 2024). Social media also has the potential to reshape dominant narratives by highlighting examples of older individuals who are active, resilient, and socially engaged (Fuente-Hernández et al., 2025).

Older age is not only overlooked but also often feared. While historically and culturally older individuals were viewed as sources of wisdom, authority and family cohesion, contemporary representations predominantly associate them with dependency, fragility, uselessness, or slowness (Camacho-Markina & Santos-Diez, 2025). This fear becomes especially pronounced in relation to the *fourth age*, which exists in the shadow of the more socially celebrated *third age*, characterized by activity and independence (Gilleard & Higgs, 2010, 2013). Future research could critically examine the extent to which HCI studies on older adults may inadvertently reinforce the fourth age by predominantly focusing on the third. Such reflection is essential for ensuring that HCI contributes to a more inclusive and balanced understanding of later life.

2.2 Ageing

This section outlines several definitions of ageing, highlighting the shared understanding that ageing is a universal process of change. Despite this common framing, it remains largely underrepresented in older-adult HCI research. The section also examines the pervasive social discourse surrounding ageing, both in broader society and within HCI. It further explores how this discourse influences the ways in which researchers engage with and design for older adults.

2.2.1 A Universal Process of Change

Ageing is a universal experience, being one of the few phenomena that connects all human beings. Yet, it is defined in diverse and sometimes conflicting ways. Common conceptualizations include healthy, active, productive, optimal, and independent ageing. At the same time, ageing has been framed as a source of pain (Lane & Smith, 2018), a biomedical issue (Kaufman et al., 2004), and a socially constructed phenomenon (Gilleard, 2025; Joyce & Loe, 2010). Ageing can also be interpreted through different temporal lenses: as biological time (e.g. wrinkles and grey hair), biographical time (e.g. life stages), and social or institutional time (e.g. retirement and grandparenthood) (Carr & Komp, 2011).

These varied understandings of ageing are reflected in HCI research. Some studies focus on promoting or supporting active and independent ageing, e.g. (Bekiaris & Bonfiglio, 2009; Karkera et al., 2023; Olatunji et al., 2021), while others explore the social dimensions of ageing and digital technology use, such as intergenerational relationships and communication, e.g. (Berridge & Wetle, 2020; Chen & Li, 2024; Wang et al., 2024).

What is ageing? The answer often depends on the perspective adopted. Nonetheless, a common thread across many definitions is that ageing fundamentally involves change (George et al., 2016). As people grow older, we accumulate years and undergo transformations across multiple dimensions. Some of these changes may be within one's control, such as quitting smoking or adopting healthier eating habits to manage cholesterol levels. Others, however, lie beyond individual control (Cruikshank, 2009; Gilleard, 2025; Lane & Smith, 2018). Research has shown that structural determinants, including governance, social and public policies, and prevailing cultural values, play a significant role in shaping an older individual's socioeconomic status, such as education, occupation, and income (Perez et al., 2022). Ageing, therefore, is not only a personal experience but also a societal and cultural phenomenon (Fung, 2013; Bengtson et al., 2016; Khanuja, 2025).

Implicit in this definition of ageing is the understanding that it is a process rather than a sudden transition. Individuals do not simply wake up one day, at age 65 or any other threshold, and become "older". Instead, ageing unfolds as a complex interplay of gains and losses over time. Despite some notable exceptions, e.g. (Brewer & Piper, 2016; Jones et al., 2023; Knowles et al., 2021; Romero et al., 2010), HCI research concerning older adults often treats ageing as a fixed state or a discrete point in time, rather than as a dynamic and evolving process.

2.2.2   A Physical, Social, and Cultural Experience

There is a prevailing tendency to view ageing primarily as a physical experience, with many of the challenges older adults face in daily life attributed to the inevitable effects of biological ageing. For instance, age-related changes in vision may lead to difficulties reading small icons on increasingly compact digital screens. As we age, functional abilities naturally decline—this is an unavoidable aspect of the human condition. However, like all human experiences, ageing is not solely biological; it is also deeply social and cultural (Cho & Gutchess, 2025; Fung, 2013; Keith, 1998; Morganroth, 2004).

Examples of the multifaceted nature of ageing can be found in everyday ageism (Sayago, 2023a, 2023b), self-stereotyping and its negative impact on well-being (Dang & Zhang, 2025), the role of experiential factors such as lifestyle in mediating cognitive ageing (Cho & Gutchess, 2025), and significant differences between countries regarding attitudes towards ageing and care as well as perceptions and acceptance of life logging technologies (Heek et al., 2020). These perspectives also extend to digital technology design. While biological ageing can contribute to interaction challenges, research suggests that many of the difficulties older adults face when using computers stem more from poor design than from declining capabilities (Sayago, 2019).

Despite its universality, there is a widespread desire to grow older without ageing. This reflects a growing denial of what is considered "normal" ageing. The biomedicalization of ageing is increasingly difficult to avoid (Kaufman et al., 2004), as societal focus shifts towards ideals of successful, productive, active, and optimal ageing (Settersten & Angel, 2011; Langmann & Weßel, 2023), or other forms of so-called positive ageing.

The social discourse surrounding ageing often employs a vocabulary, such as "handling", "looking after", "support", and "help", that implicitly frames older adults as a problem, e.g. (Bossio et al., 2023; Hazan, 1994; Huang & Davis, 2024). Although there are exceptions, youth and old age are frequently used as proxies for good and bad, respectively. Contemporary society tends to value novelty, speed, and urgency, prioritizing the present over the past. This contributes to a broader cultural narrative in which older people appear to be out of place in today's fast-paced world (Cabrera & Malanowski, 2009).

A significant portion of HCI research involving older adults reflects the dominant social discourse on ageing, often emphasizing problems and deficits. This is evident in the rhetoric of compassion, where technology is framed as a means to help older people (Rogers & Marsden, 2013). Such framing raises the old but still important question about whether the challenges being addressed truly reflect the needs of older adults or rather serve those who perceive older individuals as a burden (Hazan, 1994). However, a growing body of HCI research challenges this narrative (Lazar et al., 2025), opening up space to critically examine how different conceptualizations of ageing influence current practices and shape future directions. Chapter 3 explores this issue in greater depth.

2.3 Older People

This section explores the question of who older adults are, arguing that they can be understood as both a homogenous and heterogeneous user group, depending on the perspective adopted. It highlights that a great deal of older adults are self-aware individuals who demonstrate agency in their everyday lives. Moreover, despite age-related changes in capabilities, many older adults continue to function effectively and even exhibit creativity.

2.3.1 Who Are They?

While most people believe they know what it means to be an older adult, providing a single, universal definition remains a challenge, and such a definition may not even exist. The question of who older adults are does not lend itself to a simple or singular answer. Historically, the concept of "older adult" has encompassed both subjective and objective (i.e. official) definitions (Thane, 2005). Some individuals may be classified as older based on chronological age, typically 60 or 65, according to contemporary standards. Yet others, despite meeting these criteria, may not perceive themselves as old, may not feel old, or may actively resist being labelled as such.

For years, scientific literature has advocated for a view of older adults and later life that reflects the diversity and multidimensionality of the ageing experience (Klusmann & Kornadt, 2020). As previously discussed, older age is inherently diverse, partly due to the sheer length of this life stage. Moreover, ageing is a process of change, and this process does not unfold identically for everyone (Cho & Gutchess, 2025).

While variability is often emphasized as a defining characteristic of older adults, this focus can sometimes overshadow the commonalities they share (Settersten & Angel, 2011). For example, nearly all older individuals experience age-related changes in functional abilities and are likely to encounter age discrimination, both in everyday life and online (Sayago, 2023a, 2023b). Despite individual differences, the behaviour of many older adults, including their interactions with digital technologies, can be understood through various theories and frameworks (see Sect. 2.3.3), even if their experiences remain uniquely personal (see Fig. 2.2).

HCI research has identified several common themes in older adults' experiences with digital technologies, including their attitudes, opinions, and patterns of use. Older adults tend to perceive themselves as slow learners of digital technologies (Sayago et al., 2013; Tanprasert et al., 2024). A substantial body of evidence highlights the role of fear and concerns about technology replacing human interaction as key factors influencing technology (non)use among older adults, e.g. (Arthanat et al., 2019; Caleb-Solly et al., 2014;

Fig. 2.2 Above, older adults in Dundee (Scotland). Below, older adults in Barcelona (Spain). Socialization was a key ingredient in their learning experiences of computer use and motivations for exploring what they could do with computers and digital technologies, despite the clear and obvious socio-cultural differences between both groups. Photos by the author

Deutsch et al., 2019; Mendel & Tocj, 2019; Nimrod, 2018; Odu & Wyk, 2025; Sato et al., 2011; Tudorie, 2023; Wilson et al., 2023). The social environment, particularly the influence of relatives, instructors, partners, and friends (see Fig. 2.2), plays a decisive role in whether and how many older adults learn to use a wide range of digital technologies, from computers to automated vehicles, e.g. (Berridge & Wetle, 2020; Batbold et al., 2024; Sayago et al., 2013; Ten Bruggencate et al., 2019; Wiesent et al., 2024). Interestingly, children are often not effective teachers for their older parents (Portz et al., 2019; Sayago et al., 2013). They can even underestimate their parents' ability to comprehend the functions of the technologies and the importance of engaging them fully in decision-making (Berridge & Wetle, 2020). Across diverse profiles, older adults frequently express a strong desire to remain active and pursue personal growth, which motivates their engagement in meaningful activities mediated by digital technologies (Zhao et al., 2024). Importantly, technology acceptance among older adults is rarely an isolated, individual decision; it is typically shaped by the influence of close social circles, including children, grandchildren, friends, and neighbours (Luijkx et al., 2015; Wilson et al., 2023).

Older adults can be understood as both a highly heterogeneous and homogeneous user group. This duality depends on whether we emphasize individual variability, for example, the fact that people age in different ways, or focus on the shared experiences and challenges that unite them as a group (see Fig. 2.2). If we lose sight of these commonalities, ignore them, or fail to examine older adults through this lens, we risk overlooking the very factors that make older-adult HCI a distinct and meaningful area of research.

2.3.2 Functioning Beyond Changing Capabilities

Research consistently shows that increased age is often associated with lower performance on various cognitive tests (Cornelis et al., 2019; Salthouse, 2010). Ageing is typically accompanied by declines in processing speed, reduced efficiency in inhibiting irrelevant thoughts, and diminished capacity to manage complex tasks (Thomas & Gutchess, 2020). However, these age-related declines do not preclude the possibility of cognitive improvement across cohorts of older adults (Rehnberg et al., 2024).

The impact of cognitive decline is often moderated by familiarity with the environment. It tends to be most pronounced when older adults are placed in unfamiliar settings or asked to perform novel tasks (Park & Schwarz, 2000; Schomaker et al., 2022). For instance, differences in prospective memory—the ability to plan and remember future actions—between younger and older individuals can be partly explained by the level of abstraction of a task and the familiarity of the context in which it is performed (Menéndez-Granda et al., 2025). Importantly, age-related cognitive declines do not necessarily impair daily functioning. This is because individuals rarely operate at their maximum cognitive capacity in everyday life and can rely on experience, social support, memory aids, and other compensatory resources to adapt to changing capabilities (Salthouse, 2010).

For a long time, older adults have often been viewed as deficient versions of younger web users (Hanson, 2009). They tend to navigate more slowly from page to page, take longer to complete tasks, revisit pages more frequently, and spend more time selecting link targets. However, research shows that older and younger users follow different paths through web content and engage in distinct strategies to accomplish the same tasks (Fairweather, 2009).

These findings suggest that older adults may approach online activities differently, not necessarily less effectively, but in ways shaped by their experiences, preferences, and cognitive styles. For example, more recent works show that older adults tend to exhibit strong uncertainty avoidance when they are learning to use digital technologies because they are afraid of making mistakes—this is a lesson they have learned, and most younger adults still have not, from their life experience (Sayago et al., 2013). Older adults can also show preference for interactive guidance from instructors to demonstrate interface operations while learning automated vehicles (Batbold et al., 2024) because this is how most of them have learned previous technologies.

Differences in task completion time and success rates between older and younger users tend to be more pronounced when no alternatives are available. However, when alternatives exist, older adults are not necessarily deficient users. Studies have shown that some older individuals outperform younger adults in ill-defined tasks, where the search problem lacks clear structure (Chin et al., 2009; Scarampi et al., 2024). Moreover, no significant age differences have been found in information search tasks that use tag- or keyword-based approaches, as opposed to traditional hierarchical navigation (Pak et al., 2009). When older adults are allowed to engage with technologies in flexible ways—without rigid scripts or predefined procedures—performance differences across age groups tend to disappear. For example, an interesting study of comprehension of written sarcasm online found that when an emoji is present, the performance of older adults becomes as good as (or at least not significantly different from) that of younger adults. In the absence of an emoji, older adults, in comparison with their younger counterparts, demonstrated deficient ability in comprehending written sarcastic language (Garcia et al., 2022).

Negative stereotypes about older adults and age-related cognitive changes have long suggested that creativity tends to decline with age. However, recent evidence indicates that while cognition and creativity share functional overlaps, creativity appears to be well preserved in later life (Ross et al., 2023), even when it comes to digital technologies use (see Image 3). Rather than simply following predefined steps while navigating e-government websites, older adults can engage in a process of exploration and adaptation, experimenting with different features to complete tasks. This sense-making process is not merely functional but also creative, highlighting the ways in which older individuals actively shape their digital interactions to overcome barriers and achieve their goals (Bar-Lev et al., 2024).

Fig. 2.3 Group of older women creating questions for a digital game by using several technologies and engaging in a discussion about the topics and how to write the questions and answers to make them appealing to the players. Photo by the author

Taken together, these findings invite us to reconsider how we conceptualize older adults as users of digital technologies. Older people are often patronized in technology-related discourse, which contributes to the development of stigmatizing technologies that many older adults are reluctant to adopt (Caldeira et al., 2022; Vargemidis et al., 2021). Age-based discrimination remains prevalent (Sayago, 2023a, 2023b), with the COVID-19 pandemic serving as a stark example, leading to unwanted isolation for many older individuals (Barbosa-Neves et al., 2019). To better understand older adults within HCI, it is essential to account for changing capabilities, as interaction with digital systems relies heavily on cognitive processing. However, it is equally important to move beyond a deficit-based view of ageing (Rogers & Marsden, 2013; Sayago, 2019), recognizing that older adults continue to function—and even demonstrate creativity (see Fig. 2.3)—despite age-related changes, e.g. (Alm et al., 2009; Davidson & Jensen, 2013; Gallistl, 2018; Sayago et al., 2016; Zou et al., 2024).

2.3.3   Self-aware Individuals with Agency

In line with widespread, and often negative, views of older adults, they are frequently portrayed as passive recipients of the biological ageing process. However, by drawing on several theoretical frameworks that explain older adults' everyday behaviour, a more nuanced and empowering perspective emerges. This section focuses on three influential models: Selective Optimization with Compensation (SOC), Socioemotional Selectivity Theory (SST), and the Life-Span Theory of Control. These frameworks are widely recognized in interdisciplinary ageing research and offer valuable insights into how older adults adapt, prioritize, and maintain agency throughout later life.

The *Selective Optimization with Compensation* (SOC) theory addresses how older adults maintain functioning despite age-related declines (Baltes & Baltes, 1990). According to the SOC model, selection refers to an increasing restriction of one's life world to fewer domains of functioning because of an ageing loss. Optimization is about the maximization of gains, while compensation aims at counteracting or avoiding losses, thus ensuring the maintenance of functioning.

According to the *Socioemotional Selectivity Theory* (SST), which is a life-span theory of motivation, time horizons play a key role in motivation. When time is perceived as open-ended, goals that become most prioritized are focused on gathering information, experiencing novelty, and expanding the breadth of knowledge. When time is perceived as constrained, the most salient goals are those that can be realized in the short term. As people age, we attach less importance to goals that expand their horizons and greater importance to goals from which they derive emotional meaning (Carstensen, 2006). Older people might have clearer priorities and values, as opposed to younger adults (Carstensen, 2021).

The *Life-Span Theory of Control* (Heckhausen et al., 2010) proposes that the key criterion for adaptive development is the extent to which the individual realizes control of his or her environment. However, as individuals' capacity for primary control decreases in old age, individuals need to have strategies that facilitate disengagement from unattainable goals in favour of pursuing other more attainable ones. As with SST and SOC, this theory proposes a need to allocate diminishing resources resulting from age-related declines in sensory, motor, and cognitive abilities; strategies for compensating for losses; and methods for optimizing adaptive development (Schulz et al., 2015).

Socioemotional Selective Theory, Selective Optimization with Compensation Theory, and Life-Span Theory of Control each fall under the umbrella of the Continuity Theory of Ageing (Atchley, 1989), which contends that ageing is a dynamic and evolutionary developmental process in which individuals grow, adapt, and change. This continuity manifests itself in discussions around envisioning future design of technologies for supporting meaningful activities in later life (Zhao et al., 2024).

These theories can potentially change the way most people think about older adults. These theories allow us to realize that most older adults exhibit agency. They are also resourceful in dealing with the failings of their own body and adapting their behaviour, self-concept, and relationships with others to maintain a quality of life (Fitzpatrick et al., 2015).

Socioemotional Selective Theory, Selective Optimization with Compensation Theory, and Life-Span Theory of Control can also aid in increasing the theoretical component of HCI research with older people by explaining their technology use (or lack of) (Sayago,

2019). For example, why do some older adults use social networking sites to strengthen existing social ties rather than expand their circle of contacts? According to SST, older adults attach a great deal of importance to goals from which they derive emotional meaning (e.g. close ties). Why do most older people, especially when they are learning to use computers and the Internet, tend to exhibit a strong uncertainty avoidance in their technology use? According to the SOC model, reducing the exploration of new things (in this case, technologies and functionalities) is a functional adaptation.

These theories also offer valuable guidance for designing digital technologies for older adults. As Schulz et al. (2015) note, common threads across these frameworks include: (1) the need to allocate diminishing resources due to age-related declines in functional abilities; (2) strategies for compensating for losses and focusing on existing strengths; and (3) approaches for optimizing adaptive development. Designing technologies solely to compensate for changing abilities is insufficient. A more human-centred approach begins by recognizing older individuals as people—not just users defined by limitations, e.g. (Knowles & Hanson, 2018; Knowles et al., 2021; Lazar et al., 2016a, 2016b; Liu et al., 2016). Most older adults are aware of their strengths and weaknesses and demonstrate agency in their decisions and actions (see Fig. 2.4). For example, opinions of older adults about whether to share information with their friends, families, caregivers, and doctors have been found to be highly context-dependent and involve weighing complex trade-offs, involving benefits, recipients and risks (Frik et al., 2023). Despite acknowledging the potential mobility advantages of self-driving vehicles, many older adults express significant concerns that undermine their trust in this technology (Huff et al., 2019).

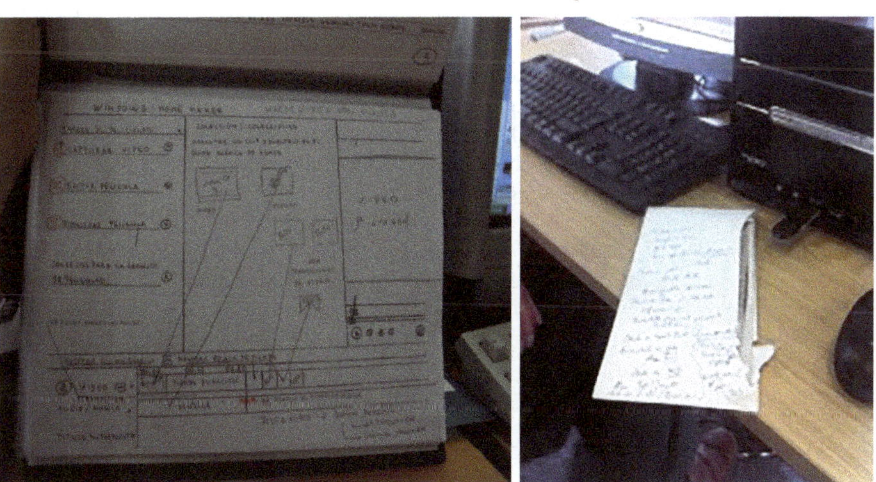

Fig. 2.4 Difficulties in remembering computer jargon and how to conduct tasks were pervasive. To overcome them, my older research participants turned to their handwritten (and digitalised) notes. Photo by the author

References

Alm, N., Astell, A., Gowans, G., Dye, R., Ellis, M., Vaughan, P., et al. (2009). Lessons learned from developing cognitive support for communication, entertainment, and creativity for older people with dementia. In C. Stephanidis (Ed.), *Universal access in human-computer interaction. Addressing diversity* (Vol. 5614, pp. 195–201). Springer Berlin Heidelberg. https://doi.org/10.1007/978-3-642-02707-9_21

Arthanat, S., Vroman, K. G., Lysack, C., & Grizzetti, J. (2019). Multi-stakeholder perspectives on information communication technology training for older adults: Implications for teaching and learning. *Disability and Rehabilitation: Assistive Technology, 14*(5), 453–461. https://doi.org/10.1080/17483107.2018.1493752

Ashcroft, A., & Knowles, B. (2025). Beyond binary: Re-imagining age using gender in HCI for inclusive design. In *Proceedings of the extended abstracts of the CHI conference on human factors in computing systems* (pp. 1–7). https://doi.org/10.1145/3706599.3719766

Atchley, R. C. (1989). A continuity theory of normal aging. *Gerontologist, 29*(2), 183–190. https://doi.org/10.1093/geront/29.2.183. PMID: 2519525.

Baltes, P. B., & Baltes, M. M. (1990). Psychological perspectives on successful aging: The model of selective optimization with compensation. In P. B. Baltes & M. M. Baltes (Eds.), Successful aging: Perspectives from the behavioral sciences (pp. 1–34). Cambridge University Press. https://doi.org/10.1017/CBO9780511665684.003

Bar-Lev, S., Aisenberg-Shafran, D., & Luria, A. (2024). Navigating e-government: Older adults' strategies and barriers in accessing state social benefits. *SSRN.* https://doi.org/10.2139/ssrn.4934321

Barbosa Neves, B., Franz, R., Judges, R., Beermann, C., & Baecker, R. (2019). Can digital technology enhance social connectedness among older adults? A feasibility study. *Journal of Applied Gerontology, 38*(1), 49–72. https://doi.org/10.1177/0733464817741369

Barrett, A. E. (2022). Centering age inequality: Developing a sociology-of-age framework. *Annual Review of Sociology, 48*(1), 213–232. https://doi.org/10.1146/annurev-soc-083121-043741

Barros Pena, B., Clarke, R. E., Holmquist, L. E., & Vines, J. (2021). Circumspect users: Older adults as critical adopters and resistors of technology. In *Proceedings of the 2021 CHI conference on human factors in computing systems* (pp. 1–14). https://doi.org/10.1145/3411764.3445128

Batbold, T., Soro, A., & Schroeter, R. (2024). Mentorable interfaces for automated vehicles: A new paradigm for designing learnable technology for older adults. In *Proceedings of the CHI conference on human factors in computing systems* (pp. 1–15). https://doi.org/10.1145/3613904.3642390

Bekiaris, E., & Bonfiglio, S. (2009). The OASIS concept. In C. Stephanidis (Ed.), *Universal access in human-computer interaction. Addressing Diversity* (Vol. 5614, pp. 202–209). Springer Berlin Heidelberg. https://doi.org/10.1007/978-3-642-02707- 9_22

Bengtson, V. L., Settersten, R., Dr., P., Settersten, R., Jr, D., & PhD. (2016). *Handbook of theories of aging*, Third Edition. Springer.

Berridge, C., & Wetle, T. F. (2020). Why older adults and their children disagree about in-home surveillance technology, sensors, and tracking. *The Gerontologist, 60*(5), 926–934. https://doi.org/10.1093/geront/gnz068

Bossio, D., McCosker, A., Schleser, M., Davis, H., & Randjelovic, I. (2023). Not that old person: Older people's responses to ageism revealed through digital storytelling. *Journal of Sociology, 59*(1), 232–250. https://doi.org/10.1177/14407833211040111

Brewer, R., & Piper, A. M. (2016). 'Tell it like it really is': A case of online content creation and sharing among older adult bloggers. In *Proceedings of the 2016 CHI conference on human factors in computing systems* (pp. 5529–5542). https://doi.org/10.1145/2858036.2858379

Cabrera, M., & Malanowski, N. (Eds.). (2009). *Information and communication technologies for active ageing: Opportunities and challenges for the European Union*. IOS Press

Caldeira, C., Nurain, N., & Connelly, K. (2022). "I hope I never need one": Unpacking stigma in aging in place technology. In *CHI conference on human factors in computing systems* (pp. 1–12). https://doi.org/10.1145/3491102.3517586

Caleb-Solly, P., Dogramadzi, S., Ellender, D., Fear, T., & Heuvel, H. V. D. (2014). A mixed-method approach to evoke creative and holistic thinking about robots in a home environment. In *Proceedings of the 2014 ACM/IEEE international conference on human-robot interaction* (pp. 374–381).

Camacho-Markina, I., & Santos-Diez, M.-T. (2025). The social construction of age: Media stigmatization of older adults: a systematic review. *Journalism and Media, 6*(3), 150. https://doi.org/10.3390/journalmedia6030150

Carr, D., & Komp, K. (2011). *Gerontology in the era of the Third Age: Implications and next steps* (Vol. 49). Springer. http://choicereviews.org/review/https://doi.org/10.5860/CHOICE.49-0892

Carstensen, L. L. (2006). The Influence of a sense of time on human development. *Science, 312*(5782), 1913–1915. https://doi.org/10.1126/science.1127488

Carstensen, L. L. (2021). Socioemotional selectivity theory: The role of perceived endings in human motivation. *The Gerontologist, 61*(8), 1188–1196. https://doi.org/10.1093/geront/gnab116

Chen, B., & Li, X. (2024). Understanding socio-technical opportunities for enhancing communication between older adults and their remote family. In *Proceedings of the CHI conference on human factors in computing systems* (pp. 1–16). https://doi.org/10.1145/3613904.3642318

Chin, J., Fu, W.-T., & Kannampallil, T. (2009). *Adaptive information search: Age-dependent interactions between cognitive profiles and strategies*.

Cho, I., & Gutchess, A. (2025). How age and culture influence cognition: A lifespan developmental perspective. *Developmental Review, 75*, Article 101169. https://doi.org/10.1016/j.dr.2024.101169

Chulián Horrillo, A., Valdivia-Salas, S., & Páez, M. (2024). Una mirada contextual al edadismo. *Análisis y Modificación de Conducta, 50*(182), 69–88. https://doi.org/10.33776/amc.v50i182.8064

Cornelis, M. C., Wang, Y., Holland, T., Agarwal, P., Weintraub, S., & Morris, M. C. (2019). Age and cognitive decline in the UK Biobank. *PLoS ONE, 14*(3), Article e0213948. https://doi.org/10.1371/journal.pone.0213948

Cruikshank, M. (2009). *Learning to be old: Gender, culture, and aging (2nd ed)*. Rowman & Littlefield Pub.

Czaja, S. J., & Weingast, S. Z. (2020). The changing face of aging: Characteristics of older adult user groups. *Gerontechnology, 19*(2), 115–124. https://doi.org/10.4017/gt.2020.19.2.004.00

Dang, Q., & Zhang, B. (2025). Sense of control and aging self-stereotyping: The role of need for structure. *BMC Psychology, 13*(1), 732. https://doi.org/10.1186/s40359-025-03061-9

Davidson, J. L., & Jensen, C. (2013). *Participatory design with older adults: An analysis of creativity in the design of mobile healthcare applications*. C&C

Deutsch, I., Erel, H., Paz, M., Ho@man, G., & Zuckerman, O. (2019). Home

Durick, J., Robertson, T., Brereton, M., Vetere, F., & Nansen, B. (2013). Dispelling ageing myths in technology design. In *Proceedings of the 25th Australian computer-human interaction conference: Augmentation, application, innovation, collaboration* (pp. 467–476). https://doi.org/10.1145/2541016.2541040

Durrant, A., Kirk, D., Trujillo Pisanty, D., Moncur, W., Orzech, K., Schofield, T., et al. (2017). Transitions in digital personhood: Online activity in early retirement. In *Proceedings of the 2017 CHI conference on human factors in computing systems* (pp. 6398–6411). https://doi.org/10.1145/302 5453.3025913

Fairweather, P. G. (2009). Influences of age and experience on web-based problem solving strategies. In C. Stephanidis (Ed.), *Universal access in human- computer interaction. Addressing Diversity* (Vol. 5614, pp. 220–229). Springer Berlin Heidelberg. https://doi.org/10.1007/978-3-642-02707-9_24

Fitzpatrick, G., Huldtgren, A., Malmborg, L., Harley, D., & Ijsselsteijn, W. (2015). Design for agency, adaptivity and reciprocity: reimagining AAL and telecare agendas. In V. Wulf, K. Schmidt, & D. Randall (Eds.), *Designing socially embedded technologies in the real-world* (pp. 305–338). Springer.

Frik, A., Bernd, J., & Egelman, S. (2023). A model of contextual factors affecting older adults' information-sharing decisions in the U.S. *ACM Transactions on Computer-Human Interaction, 30*(1), 1–48. https://doi.org/10.1145/3557888

Fuente-Hernández, L., De La Fuente-Ruiz, E., Gracia-García, P., Serrano-García, I., Álvarez-Mon, M. Á., & Molina-Ruiz, R. M. (2025). Using social media-based positive education to challenge negative stereotypes of aging: A quasi-experimental approach. *BMC Public Health, 25*(1), 533. https://doi.org/10.1186/s12889-025-21327-0

Fung, H. H. (2013). Aging in culture. *The Gerontologist, 53*(3), 369–377. https://doi.org/10.1093/geront/gnt024

Gallistl, V. (2018). The emergence of the creative ager—On subject cultures of late-life creativity. *Journal of Aging Studies, 46*, 93–99. https://doi.org/10.1016/j.jaging.2018.07.002

Garcia, C., Ţurcan, A., Howman, H., & Filik, R. (2022). Emoji as a tool to aid the comprehension of written sarcasm: Evidence from younger and older adults. *Computers in Human Behavior, 126*, Article 106971. https://doi.org/10.1016/j.chb.2021.106971

George, L. K., Ferraro, K. F., Carr, D., Wilmoth, J. M., & Wolf, D. (Eds.). (2016). *Handbook of aging and the social sciences (Eighth edition)*. Elsevier.

Ghorayeb, A., Comber, R., & Gooberman-Hill, R. (2021). Older adults' perspectives of smart home technology: Are we developing the technology that older people want? *International Journal of Human-Computer Studies, 147*, Article 102571. https://doi.org/10.1016/j.ijhcs.2020.102571

Gilleard, C. (2025). Revisiting the social construction of old age. *Ageing and Society, 45*(3), 500–513. https://doi.org/10.1017/S0144686X23000570

Gilleard, C., & Higgs, P. (2010). Aging without agency: Theorizing the fourth age. *Aging & Mental Health, 14*(2), 121–128. https://doi.org/10.1080/13607860903228762

Gilleard, C., & Higgs, P. (2013). The fourth age and the concept of a 'social imaginary': A theoretical excursus. *Journal of Aging Studies, 27*(4), 368–376. https://doi.org/10.1016/j.jaging.2013.08.004

Guner, H., & Acarturk, C. (2020). The use and acceptance of ICT by senior citizens: A comparison of technology acceptance model (TAM) for elderly and young adults. *Universal Access in the Information Society, 19*(2), 311–330. https://doi.org/10.1007/s10209-018-0642-4

Hanson, V. L. (2009). Cognition, age, and web browsing. In C. Stephanidis (Ed.), Universal access in human-computer interaction. Addressing diversity (Vol. 5614, pp. 245–250). Springer Berlin Heidelberg. https://doi.org/10.1007/978-3-642-02707-9_27

Hazan, H. (1994). *Old age constructions and deconstructions*. Cambridge University Press.

Heckhausen, J., Wrosch, C., & Schulz, R. (2010). A motivational theory of life-span development. *Psychological Review, 117*(1), 32–60. https://doi.org/10.1037/a0017668

Heek, J., Wilkowska, W., & Ziefle, M. (2020). Colors of aging: Cross-cultural perception of lifelogging technologies in older age. In *Proceedings of the 6th international conference on information and communication technologies for ageing well and E-health* (pp. 38–49). https://doi.org/10.5220/0009371000380049

Huang, L., & Davis, B. (2024). *Language, Aging, and Society, What can linguistics do for the aging world?* Palgrave macmillan.

Huff, E. W., DellaMaria, N., Posadas, B., & Brinkley, J. (2019). Am i too old to drive?: Opinions of older adults on self-driving vehicles. In *Proceedings of the 21st international ACM SIGACCESS conference on computers and accessibility* (pp. 500–509). https://doi.org/10.1145/3308561.335 3801

Johfre, S., & Saperstein, A. (2023). The social construction of age: Concepts and measurement. *Annual Review of Sociology, 49*(1), 339–358. https://doi.org/10.1146/annurev-soc-031021-121020

Jones, W. P., Donner, S., Narayan, B., & Reyes, V. (2023). It's about Time: Let's Do More to Support the Process of Aging (vs. the State of Being "Old"). In *Extended abstracts of the 2023 CHI conference on human factors in computing systems* (pp. 1–7). https://doi.org/10.1145/3544549.3582740

Joyce, K., & Loe, M. (2010). *Technogenarians. Studying health and ilness though an ageing, science, and technology lens.* Wiley-Blackwell.

Kannabiran, G., Hoggan, E., & Koefoed Hansen, L. (2020). *Somehow they are never horny! companion publication of the 2020 ACM designing interactive systems conference* (pp. 131–137). https://doi.org/10.1145/3393914.3395877

Karkera, Y., Tandukar, B., Chandra, S., & Martin-Hammond, A. (2023). Building community capacity: Exploring voice assistants to support older adults in an independent living community. In *Proceedings of the 2023 CHI conference on human factors in computing systems* (pp. 1–17). https://doi.org/10.1145/3544548.3581561

Kaufman, S. R., Shim, J. K., & Russ, A. J. (2004). Revisiting the biomedicalization of aging: Clinical trends and ethical challenges. *The Gerontologist, 44*(6), 731–738. https://doi.org/10.1093/geront/44.6.731

Keith, J. (Ed.). (1998). *The aging experience: Diversity and commonality across cultures (Nachdr.).* SAGE.

Khanuja, N. (2025). Designing technological interventions to enhance self-perceptions of aging (SPA) and well-being of older adults in India. In *Designing Interactive Systems Conference (DIS '25 Companion)*, July 05–09, 2025, Funchal, Portugal. ACM, New York, NY, USA. https://doi.org/10.1145/3715668.3735620

Klusmann, V., & Kornadt, A. E. (2020). Current directions in views on ageing. *European Journal of Ageing, 17*(4), 383–386. https://doi.org/10.1007/s10433-020-00585-4

Knowles, B., & Hanson, V. L. (2018). The wisdom of older technology (non)users. *Communications of the ACM, 61*(3), 72–77. https://doi.org/10.1145/3179995

Knowles, B., Hanson, V. L., Rogers, Y., Piper, A. M., Waycott, J., Davies, N., Ambe, A. H., Brewer, R. N., Chattopadhyay, D., Dee, M., Frohlich, D., Gutierrez- Lopez, M., Jelen, B., Lazar, A., Nielek, R., Pena, B. B., Roper, A., Schlager, M., Schulte, B., & Yuan, I. Y. (2021). The harm in conflating aging with accessibility. *Communications of the ACM, 64*(7), 66–71. https://doi.org/10.1145/3431280

Lane, P., & Smith, D. (2018). Culture, ageing and the construction of pain. *Geriatrics, 3*(3), 40. https://doi.org/10.3390/geriatrics3030040

Langmann, E., & Weßel, M. (2023). Leaving no one behind: Successful ageing at the intersection of ageism and ableism. *Philosophy, Ethics, and Humanities in Medicine, 18*(1), 22. https://doi.org/10.1186/s13010-023-00150-8

Lazar, A., Brewer, R. N., & Knowles, B. (2025). HCI and older adults: The critical turn and what comes next. *Foundations and Trends® in Human-Computer Interaction, 19*(2), 112–212. https://doi.org/10.1561/1100000094

Lazar, A., Cornejo, R., Edasis, C., & Piper, A. M. (2016a). Designing for the third hand: Empowering older adults with cognitive impairment through creating and sharing. In: *Proceedings of the 2016 ACM conference on designing interactive systems* (pp. 1047–1058). https://doi.org/10.1145/290 1790.2901854

Lazar, A., Thompson, H. J., Piper, A. M., & Demiris, G. (2016b). Rethinking the design of robotic pets for older adults. In *Proceedings of the 2016 ACM conference on designing interactive systems* (pp. 1034–1046). https://doi.org/10.1145/2901790.2901811

Lee, C., FakhrHosseini, M., Miller, J., Patskanick, T. R., & Coughlin, J. F. (2019). The oldest olds' perceptions of social robots. In J. Zhou & G. Salvendy (Eds.), Human aspects of IT for the aged population. Social media, games and assistive environments (Vol. 11593, pp. 405–415). Springer International Publishing. https://doi.org/10.1007/978-3-030-22015-0_32

Leonardi, C., Mennecozzi, C., Not, E., Pianesi, F., Zancanaro, M., Gennai, F., & Cristoforetti, A. (2009). Knocking on elders' door: Investigating the functional and emotional geography of their domestic space. In: *Proceedings of the SIGCHI conference on human factors in computing systems* (pp. 1703–1712). https://doi.org/10.1145/1518701.1518963

Li, J., & Waycott, J. (2023). "I feel like I'm getting younger": Investigating the social connections of older handicraft vloggers on the social platform Bilibili. In *Proceedings of the 35th Australian computer-human interaction conference* (pp. 257–265). https://doi.org/10.1145/3638380. 3638381

Lindley, S., & Wallace, J. (2015). Placing in age: Transitioning to a new home in later life. *ACM Transactions on Computer-Human Interaction, 22*(4), 1–39. https://doi.org/10.1145/2755562

Liu, N., Purao, S., & Tan, H.-P. (2016). Value-inspired service design in elderly home-monitoring systems. *IEEE International Conference on Pervasive Computing and Communication Workshops (PerCom Workshops), 2016*, 1–6. https://doi.org/10.1109/PERCOMW.2016.7457138

Luijkx, K., Peek, S., & Wouters, E. (2015). "Grandma, you should do it—it's cool" older adults and the role of family members in their acceptance of technology. *International Journal of Environmental Research and Public Health, 12*(12), 15470–15485. https://doi.org/10.3390/ijerph121 214999

Macedo, I. M. (2017). Predicting the acceptance and use of information and communication technology by older adults: An empirical examination of the revised UTAUT2. *Computers in Human Behavior, 75*, 935–948. https://doi.org/10.1016/j.chb.2017.06.013

Mendel, T., & Toch, E. (2019). My mom was getting this popup: Understanding motivations and processes in helping older relatives with mobile security and privacy. *Proceedings of the ACM on Interactive, Mobile, Wearable and Ubiquitous Technologies, 3*(4), 1–20. https://doi.org/10.1145/ 3369821

Menéndez-Granda, M., Schmidt, N., Laera, G., Clenin, A., Kliegel, M., Orth, M., & Peter, J. (2025). Factors explaining age-related prospective memory performance differences: A meta-analysis. *The Journals of Gerontology, Series B: Psychological Sciences and Social Sciences, 80*(6), gbaf020. https://doi.org/10.1093/geronb/gbaf020

Morganroth, M. (2004). *Aged by Culture.* The University of Chicago Press.

Nicholson, J., Morrison, B., Dixon, M., Holt, J., Coventry, L., & McGlasson, J. (2021). Training and embedding cybersecurity guardians in older communities. In *Proceedings of the 2021 CHI conference on human factors in computing systems* (pp. 1–15). https://doi.org/10.1145/3411764. 3445078

Nimrod, G. (2018). Technophobia among older Internet users. *Educational Gerontology, 44*(2–3), 148–162. https://doi.org/10.1080/03601277.2018.1428145

Odu, A. O., & Wyk, B. V. (2025). *Digital competencies among the greying population: A scoping review.*

Olatunji, S., Oron-Gilad, T., Markfeld, N., Gutman, D., Sarne-Fleischmann, V., & Edan, Y. (2021). Levels of automation and transparency: Interaction design considerations in assistive robots for older adults. *IEEE Transactions on Human- Machine Systems, 51*(6), 673–683. https://doi.org/10.1109/THMS.2021.3107516

Pak, R., Price, M. M., & Thatcher, J. (2009). Age-sensitive design of online health information: Comparative usability study. *Journal of Medical Internet Research, 11*(4), Article e45. https://doi.org/10.2196/jmir.1220

Park, D., & Schwarz, N. (2000). *Cognitive aging.* Taylor & Francis.

Perez, F. P., Perez, C. A., & Chumbiauca, M. N. (2022). Insights into the social determinants of health in older adults. *Journal of Biomedical Science and Engineering, 15*(11), 261–268. https://doi.org/10.4236/jbise.2022.1511023

Poortman, A.-R., & Van Tilburg, T. G. (2005). Past experiences and older adults' attitudes: A life-course perspective. *Ageing and Society, 25*(01), 19–39. https://doi.org/10.1017/S014468X04002557

Portz, J. D., Fruhauf, C., Bull, S., Boxer, R. S., Bekelman, D. B., Casillas, A., et al. (2019). "Call a Teenager... That's What I Do!"-grandchildren help older adults use new technologies: Qualitative Study. *JMIR Aging, 2*(1), e13713. https://doi.org/10.2196/13713

Rehnberg, J., Fors, S., Ford, K. J., & Leist, A. K. (2024). Cognitive performance trends among European older adults: Exploring variations across cohorts, gender, and educational levels (2007–2017). *BMC Public Health, 24*(1), 1646. https://doi.org/10.1186/s12889-024-19123-3

Renaud, K., & van Biljon, J. (2008). Predicting technology acceptance and adoption by the elderly: A qualitative study robotic devices for older adults: Opportunities and concerns. *Computers in Human Behavior, 98*, 122–133. https://doi.org/10.1016/j.chb.2019.04.002

Rogers, Y., & Marsden, G. (2013). Does he take sugar?: Moving beyond the rhetoric of compassion. *Interactions, 20*(4), 48–57. https://doi.org/10.1145/2486227.2486238

Romero, N., Sturm, J., Bekker, T., De Valk, L., & Kruitwagen, S. (2010). Playful persuasion to support older adults' social and physical activities. *Interacting with Computers, 22*(6), 485–495. https://doi.org/10.1016/j.intcom.2010.08.006

Rosales, A., & Fernández-Ardèvol, M. (2016). Beyond whatsapp: Older people and smartphones. *Romanian Journal of Communication and Public Relations, 18*(1), 27. https://doi.org/10.21018/rjcpr.2016.1.200

Ross, S. D., Lachmann, T., Jaarsveld, S., Riedel-Heller, S. G., & Rodriguez, F. S. (2023). Creativity across the lifespan: Changes with age and with dementia. *BMC Geriatrics, 23*(1), 160. https://doi.org/10.1186/s12877-023-03825-1

Rothermund, K., & De Paula Couto, M. C. P. (2024). Age stereotypes: Dimensions, origins, and consequences. *Current Opinion in Psychology, 55*, Article 101747. https://doi.org/10.1016/j.copsyc.2023.101747

Salthouse, T. A. (2010). *Major issues in cognitive aging.* Oxford University Press.

Sato, D., Kobayashi, M., Takagi, H., Asakawa, C., & Tanaka, J. (2011). How voice augmentation supports elderly web users. In *The proceedings of the 13th international ACM SIGACCESS conference on computers and accessibility* (pp. 155–162). https://doi.org/10.1145/2049536.2049565

Sayago, S (2023a). Cultures in human-computer interaction. In *Synthesis lectures on human-centered informatics.* Springer Cham

Sayago, S (2023b). Human-computer interaction research on ageism: Essential, incipient, and challenging. In A. Rosales, M. Fernández-Ardèvol & J. Svensson (Eds.), *Digital ageism. How it operates and approaches to tackling it. Routledge Studies in New Media and Cyberculture* (pp. 116–134). ISBN: 978–1–003–32368–6

Sayago, S., Forbes, P., & Blat, J. (2013). Older people becoming successful ICT learners over time: Challenges and strategies through an ethnographical lens. *Educational Gerontology, 39*(7), 527–544.

Sayago, S. (2019). *Perspectives on human-computer interaction research with older people.* Springer Cham.

Sayago, S., Rosales, A., Righi, V., Ferreira, S., Coleman, G., & Blat, J. (2016). On the conceptualization, design and evaluation of appealing, meaningful and playable digital games for older people. *Games and Culture, 11*(1–2), 53–80.

Scarampi, C., Cauvin, S., Moulin, C. J. A., Souchay, C., Schnitzspahn, K. M., Ballhausen, N., & Kliegel, M. (2024). Age- and task-setting-related performance predictions in prospective memory: Can metacognition explain the age-prospective memory paradox? *Cortex, 181*, 119–132. https://doi.org/10.1016/j.cortex.2024.09.014

Schaie, K. W., & Willis, S. L. (2011). *Handbook of the psychology of aging (7th ed).* Elsevier/Academic Press.

Schaie, K. W., Willis, S. L., Knight, B. G., Levy, B., & Park, D. C. (Eds.). (2016). *Handbook of the psychology of aging (Eighth edition).* Elsevier/Academic Press.

Schomaker, J., Baumann, V., & Ruitenberg, M. F. L. (2022). Effects of exploring a novel environment on memory across the lifespan. *Scientific Reports, 12*(1), 16631. https://doi.org/10.1038/s41598-022-20562-4

Schulz, R., Wahl, H.-W., Matthews, J. T., De Vito Dabbs, A., Beach, S. R., & Czaja, S. J. (2015). Advancing the aging and technology agenda in gerontology. *The Gerontologist, 55*(5), 724–734. https://doi.org/10.1093/geront/gnu071

Settersten, R. A., & Angel, J. L. (Eds.). (2011). *Handbook of sociology of aging.* Springer New York. https://doi.org/10.1007/978-1-4419-7374-0

Smith, K. (2025). *Touching experiences: How older adults envision ambient and tangible social technology through the lens of time.* CHI, Yokohama Japan.

Stark, J (2020). *The cult of youth. Anti-ageing in modern Britain.* Cambridge University Press.

Tang, X., Ding, X. (Sharon), & Zhou, Z. (2023). Towards equitable online participation: A case of older adult content creators' role transition on short-form video sharing platforms. *Proceedings of the ACM on Human-Computer Interaction, 7*(CSCW2), 1–22. https://doi.org/10.1145/3610216

Tanprasert, T., Dai, J., & McGrenere, J. (2024). HelpCall: Designing informal technology assistance for older adults via videoconferencing. In *Proceedings of the CHI conference on human factors in computing systems* (pp. 1–23). https://doi.org/10.1145/3613904.3642938

Ten Bruggencate, T., Luijkx, K. G., & Sturm, J. (2019). Friends or frenemies? The role of social technology in the lives of older people. *International Journal of Environmental Research and Public Health, 16*(24), 4969. https://doi.org/10.3390/ijerph16244969

Thane, P. (2005). *The long history of old age.* Thames & Hudson.

Thomas, A. K., & Gutchess, A. (Eds.). (2020). *The cambridge handbook of cognitive aging: A life course perspective (1st ed.).* Cambridge University Press. https://doi.org/10.1017/9781108552684

Thornton, J. E. (2002). Myths of aging or ageist stereotypes. *Educational Gerontology, 28*(4), 301–312. https://doi.org/10.1080/036012702753590415

Tudorie, G. (2023). Reluctant republic: A positive right for older people to refuse AI-based technology. *Societies, 13*(12), 248. https://doi.org/10.3390/soc13120248

Vargemidis, D., Gerling, K., Abeele, V. V., Geurts, L., & Spiel, K. (2021). Irrelevant gadgets or a source of worry: Exploring wearable activity trackers with older adults. *ACM Transactions on Accessible Computing, 14*(3), 1–28. https://doi.org/10.1145/3473463

Victor, C. R. (1994). *Old age in modern society.* Springer US. https://doi.org/10.1007/978-1-4899-3075-0

Von Humboldt, S., Low, G., & Leal, I. (2024). What really matters in old age? A study of older adults' perspectives on challenging old age representations. *Social Sciences, 13*(11), 565. https://doi.org/10.3390/socsci13110565

Wang, Q., & Sun, X. (2016). Investigating gameplay intention of the elderly using an Extended Technology Acceptance Model (ETAM). *Technological Forecasting and Social Change, 107*, 59–68. https://doi.org/10.1016/j.techfore.2015.10.024

Wang, S., Carmeline, A., Kolko, B., & Munson, S. A. (2024). Understanding the role of technology in older adults' changing social support networks. *Proceedings of the ACM on Human-Computer Interaction, 8*(CSCW1), 1–28. https://doi.org/10.1145/3641027

Waycott, J., Vetere, F., Pedell, S., Kulik, L., Ozanne, E., Gruner, A., et al. (2013). Older adults as digital content producers. In *Proceedings of the SIGCHI conference on human factors in computing systems* (pp. 39–48). https://doi.org/10.1145/2470654.2470662

Wiesent, T., Alnatour, J., & Seeberger, B. (2024). Smartphone use by older people in German-speaking countries: A qualitative systematic literature review. *Gerontechnology, 23*(1), 1–17. https://doi.org/10.4017/gt.2024.23.1.850.05

Willatt, A., Gray, S. I., Manchester, H., Foster, T., & Cater, K. (2024). An intersectional lifecourse lens and participatory methods as the foundations for co-designing with and for minoritised older adults. *Proceedings of the ACM on Human-Computer Interaction, 8*(CSCW1), 1–29. https://doi.org/10.1145/3637295

Wilson, G., Gates, J. R., Vijaykumar, S., & Morgan, D. J. (2023). Understanding older adults' use of social technology and the factors influencing use. *Ageing and Society, 43*(1), 222–245. https://doi.org/10.1017/S0144686X21000490

Yang, M., & Moffatt, K. (2024). Navigating the maze of routine disruption: Exploring how older adults living alone navigate barriers to establishing and maintaining physical activity habits. In *Proceedings of the CHI conference on human factors in computing systems* (pp. 1–15). https://doi.org/10.1145/3613904.3642842

Zhang, Y., Suhaimi, N., Azghandi, R., Joseph, M. A., Kim, M., Gri@in, J., et al. (2020). Understanding the use of crisis informatics technology among older adults. In: *Proceedings of the 2020 CHI conference on human factors in computing systems* (pp. 1–13). https://doi.org/10.1145/3313831.3376862

Zhao, W., Kelly, R. M., Rogerson, M. J., & Waycott, J. (2024). Older adults imagining future technologies in participatory design workshops: Supporting continuity in the pursuit of meaningful activities. In *Proceedings of the CHI conference on human factors in computing systems* (pp. 1–18). https://doi.org/10.1145/3613904.3641887

Zou, Y., Sun, K., Afnan, T., Abu-Salma, R., Brewer, R., & Schaub, F. (2024). Cross-contextual examination of older adults' privacy concerns, behaviors, and vulnerabilities. *Proceedings on Privacy Enhancing Technologies, 2024*(1), 133–150. https://doi.org/10.56553/popets-2024-0009

Ways of Older-Adult Centred Design

3

Abstract

This chapter offers a reflexive examination of older-adult centred design, situating it within the broader context of Human–Computer Interaction and the diverse, often contested perspectives on technology design for older people. It explores different ways of understanding older adults in the context of digital technology design, examines several design approaches, and outlines distinct ways of understanding technology in this field. This discussion draws on research contributions that have been published over the last 30 years.

Keywords

Older adults · Design approaches · Assumptions · Technology

3.1 Context and Purpose

A great deal of older-adult HCI research seeks to directly or indirectly enhance the design of digital technologies for older adults. The purpose of this chapter is not to provide guidance on designing a specific technology. Addressing such a question would require a different kind of inquiry. Instead, this chapter aims to offer a reflexive perspective on how we think about and approach design for this population.

3.1.1 Technology, Ageing, and Older People

Digital technology design and older adults is a subject of ongoing debate. Given that older people are said to be uninterested in digital technologies or lack the ability to use them

(Durick et al., 2013), exploring this topic can be regarded as irrelevant. Widespread (and negative) stereotypes of older adults reinforce this view (see Chap. 2). Digital technology design for older adults can also be considered too broad or unengaging, as the answer often depends on the specific technology in question. If we assume that older adults are not fundamentally different from other users, and that established human-centred and participatory design approaches are sufficient, it follows then that digital technology design for older adults is not such an important research area. It can also be argued that technology design and ageing is daunting, due to the limited understanding of older adults (Petrie, 2023), their diversity as a user group, and the complexity this introduces into the design process. These varying perspectives are examined throughout the chapter.

3.1.2 HCI and Older-Adult Design

This chapter examines the relationship between technology design and older adults through the lens of the evolution of Human–Computer Interaction (HCI) (Ashby et al., 2019; Bannon, 1991; Bødker, 2015; Frauenberger, 2020; Harrison et al., 2011; Hassenzahl, 2010; Sharma et al., 2024; Shneiderman, 2022; Xu et al., 2023)—from a focus on human factors to human actors, and ultimately to user experience, entanglement, post-growth, and human-centred AI. These shifts reflect broader changes in digital technologies, which have evolved from workplace tools used for specific tasks to ubiquitous elements of everyday life, and the "increasingly intimate entanglement of humans and digital technology in all aspects of life" (Frauenberger, 2020). They also mirror transformations in (a) user demographics, from expert users to a wider population of non-experts; (b) technology use, which have expanded beyond productivity and efficiency to include creativity, well-being, and personal meaning, and (c) human interactions with AI systems. Global crises in the making, such as rising temperatures, species extinction and degradation of ecosystems, also play an important role in this constant evolution of HCI.

Design approaches within HCI have evolved significantly over time. Human-centred design (HCD), for instance, has emerged as a dominant paradigm, largely in response to the challenges users face with existing technologies (Norman, 2005). A core tenet of HCD and HCI is the principle of "knowing your user". One of the most effective ways to achieve this is through close collaboration with them. Participatory design (PD) builds on this idea by emphasizing the active involvement of users in the design process. PD is grounded in principles of mutual learning, democratic engagement, and user empowerment (Bødker, 2022), with designers and users working together towards shared goals through reflective and inclusive practices.

While human-centred design (HCD) primarily focuses on problem-solving, alternative approaches to design offer broader possibilities. One such approach is speculative design, which uses design as a tool to imagine how things could be. It is concerned with exploring possible futures as a means to better understand the present and to provoke discussion about the kinds of futures people desire, or wish to avoid (Dunee & Raby, 2013). Another approach is Value-Sensitive Design (Friedman & Hendry, 2019), which emphasizes the importance of human values in the design process, and once the designed artefact has left the designer to be out there in the world (Ghoshal & Dasgupta, 2023). VSD focuses on ethics and morality, aiming to ensure that technologies align with what people consider meaningful and important in their lives.

Much of contemporary design practice has developed within frameworks that prioritize human perspectives and needs. As such, the ways in which we involve people in design warrants research attention, e.g. (Lee et al., 2022; Mannheim et al., 2025; Vines et al., 2013). However, more-than-human design challenges this anthropocentric orientation by recognizing that human society exists within a complex ecosystem of interactions involving both human and non-human actors (Wiberg & Teigland, 2024). This approach encourages designers to consider the agency and perspectives of non-human entities, such as animals, plants, technologies, and environments, without reinforcing human dominance over them (Eriksson et al., 2024; Giaccardi et al., 2025; Wakkary, 2021). Norman, in a recent book, *Design for a Better World* (2023), argues that "Today, that focus must expand to encompass humanity, which means all living things, the world's ecosystems, and the complex system in which we all live" (p. 100).

Concepts such as post-userism (Baumer & Brubaker, 2017) further critique the centrality of the "user" in HCI, proposing alternative frameworks that move beyond human-centric ideologies. Similarly, the notion of products as agents (Cila et al., 2017) invites designers to view networked products not merely as tools, but as entities with distributed agency, capable of acting within socio-technical systems. The boundaries between technology and humans are increasingly fuzzy, with augmented reality or neuro-implants, among other technological developments, probing the limits of where the human ends and technology starts (Frauenberger, 2020). Also, the environmental footprints of human-centred technologies are contributing to environmental problems and the climate crisis (Wiberg & Teigland, 2024; Sharma et al., 2024). These perspectives collectively expand the scope of design to include ethical, ecological, and relational dimensions that transcend traditional human-centred paradigms.

Older-adult centred design may not have advanced at the same pace as broader HCI design, in part due to its relatively recent emergence as a distinct area of inquiry.[1] To fulfil the aims of this chapter, and the wider objectives of this book, it is valuable to reflect on the current position of older-adult centred design within the broader trajectory of HCI.

---

[1] According to (Sayago, 2019), it was not until the 2000s that HCI research concerning older adults gained considerable strength, and this was thanks to the seminal works of A. Newell, P. Gregor, A. Dickinson, A. Carmichael, D. Hawthorn, M. Zajicek, S. Kurniawan, V. Hanson, and J. Goodman, amongst others. Also, the results of a bibliometric analysis of almost 200 documents published in the Web of Science and Scopus databases until 2018 show that the topic of older people and digital technologies is young and incipient, with two-thirds of the scientific production (not only in HCI) concentrated on the 2012–2018 period (Álvarez-García et al., 2019).

Such reflection can help identify gaps, opportunities, and directions for future research and practice (see Chap. 5).

3.2   Ways of Conceptualizing and Representing Older Adults

When engaging in technology design, we inevitably conceptualize the users of the technologies we create. These conceptualizations are reflected in the academic literature, shaping how different user groups are represented. This section examines how older adults are portrayed within the field of technology design. Specifically, it explores the ways in which older adults are understood, framed, and represented in scholarly discourse, and considers the implications of these representations for design practice and research.

3.2.1   Human Factors

Older adults are frequently conceptualized in technology design as individuals in need of assistance. In the academic literature, they are often represented through the lens of age-related changes in functional capabilities that require compensation. For example, research has examined the impact of declining abilities on various aspects of interaction, including user interface design (Hawthorn, 2000) and alternative methods for helping older adults to select small screen elements (Moffatt & McGrenere, 2010) and locate the mouse cursor on digital displays (Hollinworth & Hwang, 2011a, 2011b). Other studies have explored online information-seeking strategies (Chin et al., 2009; Sharit et al., 2008), compared performance between older and younger adults (Chin et al., 2009; Chin & Fu, 2012; Wu et al., 2024; While et al., 2024), and investigated cognitive training interventions (Du et al., 2024). Many of these studies are conducted in laboratory or controlled settings, where changing capabilities are a central focus.

In addition, older adults are often portrayed as passive, individual users of digital technologies—technologies that are presumed to offer solutions to a wide range of problems, from cognitive decline to health monitoring. This perspective tends to reduce older adults to human factors within the design process (Lazar et al., 2025; Sayago, 2019), overlooking their agency and lived experiences.

3.2.2   Human and Social Actors

Older adults are also increasingly recognized as active participants in their social networks and discerning users of digital technologies (see Fig. 3.1). Within this conceptualization, they are portrayed not merely as recipients of technological solutions but as contributors to digital culture and innovation. For instance, older adults have been shown to engage in

Fig. 3.1 Four older women are using a tablet to answer a geo-localised questions in an online game. They are having an active discussion on where exactly the question should be placed on the map. Photo by the author

digital content creation (Brewer & Piper, 2016; Ng & Indran, 2022; Tang et al., 2023), interact with tangible interfaces (Rebola & Jones, 2013; Rodríguez et al., 2019; Rogers et al., 2014), and even develop their own applications through programming (Guo, 2017; Sayago & Bergantiños, 2021).

Research has also explored how older adults learn about digital technologies (Mori & Harada, 2010; Pang et al., 2021; Sayago et al., 2013), maintain social connections using both mainstream and emerging platforms (Cornejo et al., 2010), contribute to their communities (Suhaimi et al., 2022; Wiles & Jayasinha, 2013), and express their preferences and critiques regarding technology (Leong & Robertson, 2016; Mitzner et al., 2010). Additionally, studies have examined their intentional non-use of digital technologies (Knowles & Hanson, 2018a, 2018b; Neven, 2010), highlighting the nuanced and context-dependent nature of their engagement.

Most of the research reflecting this perspective tends to be conducted in real-world settings or combines in-lab and out-of-lab activities, offering a more holistic view of older adults.

3.2.3 Critical Corner

Remarkably different ways of conceptualizing the same user group—older adults—coexist within the literature,[2] and this diversity can give rise to tensions in the field. For instance, researchers who work with older adults in care facilities, where individuals often experience significant age-related declines and high levels of dependency, may view older adults primarily through the lens of human factors. From this perspective, portraying older adults as active agents with autonomy might seem outside the scope of their work, or even unrealistic. Conversely, researchers who engage with older adults in community centres, neighbourhood activities, or independent living contexts may find a deficit-focused view too narrow and inconsistent with their observations.

Importantly, neither perspective is inherently incorrect; they simply reflect different contexts, research experiences and the diversity encompassed by the term "older people". In line with recommended practices for evaluating and reporting qualitative research in HCI (Soden et al., 2024), including author positionality and detailed participant profiles that go beyond chronological age, it might be valuable for studies involving older adults to explicitly discuss these contextual factors. Doing so can help prevent misunderstandings, foster more nuanced interpretations, and ultimately strengthen the quality and relevance of research in this field.

Divergent representations and conceptualizations of older adults have important implications for the evolution of the field. Early HCI studies concerning older adults—particularly those published in the late 1990s and the early 2000s, or even earlier within Human Factors and Ergonomics—primarily focused on age-related changes in functional abilities (Sayago, 2019). Over time, however, the field has shifted from viewing older adults as a set of human factors to recognizing them as social beings with agency. What might have prompted this change?

This evolution likely reflects broader societal transformations in how ageing is experienced and understood. It may also be influenced by shifts within HCI itself, such as the move from human factors to human actors (Bannon, 1991), and from usability to user experience (Hassenzahl, 2010). These changes have expanded the scope of design to include emotional, social, and experiential dimensions, aligning more closely with the lived realities of many older adults. As such, older-adult centred design remains a moving target, like HCI, which is continually evolving in response to both technological developments and changing societal understandings of ageing.

---

[2] Other conceptualisations can also be found in the literature, such as consumers/creators of digital content (Brewer & Piper, 2016), and individuals living in frailty (Burema, 2022; Rogers & Marsden, 2013). I believe that these and other views available in the literature fit in well with the main two types outlined in this section.

3.3   Core Design Assumptions

Technology design for older adults is often shaped by four core assumptions: chronological age, special needs, simplicity, and solutionism. This section outlines each of these assumptions and discusses their implications for design practice and research.

3.3.1   Chronological Age

One of the first considerations when designing for older adults is identifying who they are. This identification often relies heavily on chronological age. However, simply stating that someone is aged 60 or 65 + is insufficient for designing accessible, usable, and meaningful digital technologies that older adults genuinely want to use. Older adults are not merely individuals who have lived longer than most of us; they are gendered individuals (Ashcroft & Knowles, 2025) with agency (Knowles & Hanson, 2018b), a wealth of life experience (Leong & Robertson, 2016; Righi et al., 2017), personal interests and abilities (Guo, 2017; Sayago & Bergantiños, 2021), and, of course, age-related changes in functional abilities (Fisk et al., 2009; McLaughlin & Pak, 2020).

Relying solely on chronological age—without specifying an age range (e.g. 65–75)—reinforces the assumption that older adults form a homogeneous group. This oversimplification overlooks the significant variability in later life (Li et al., 2021) and contributes to otherism, where older adults are perceived as fundamentally different from the rest of the population. Ultimately, this can lead to both ageism and the belief that all older people are alike, which risks marginalizing their diverse experiences, capabilities, and preferences in technology design.

3.3.2   Simplicity

It is commonly assumed that older adults do not need complex technologies. Instead, reduced or limited functionality, enabling basic tasks to be performed in a straightforward manner, is often seen as sufficient. This assumption is reinforced by stereotyped views of older adults (Hummert, 2011), which have become increasingly negative over time (Ng et al., 2015). Age-related changes in functional abilities, particularly in cognitive domains such as fluid intelligence (McLaughlin & Pak, 2020; Park & Schwarz, 2000), further support the belief that older adults may need simpler technologies.

The emphasis on simplicity is not only externally imposed; it may also reflect older adults' own preferences (Tanprasert et al., 2024). For example, Zhao et al. (2024) argue that older participants in participatory design workshops expressed a desire for technologies that are simple, positive, proactive, and well-integrated into their lives. While simplicity can enhance accessibility, over-reliance on this assumption risks underestimating older adults' capabilities and interests, potentially leading to patronizing or overly constrained design solutions.

The simplicity criterion has influenced the design of various technologies for older adults, including mobile phones (Massimi et al., 2007), sympathetic devices (Rebola & Jones, 2013), ambient and tangible systems (Smith, 2025), virtual reality platforms (Abeele et al., 2021), and assistive technologies (Maskeliūnas et al., 2019). However, this raises a critical question: do older adults truly need simple technologies?

Abeele et al. (2021) point out that simplistic virtual reality (VR) environments may not offer sufficient challenge and depth for broad audiences of older people, who can successfully engage with current commercial VR devices. In a survey of mobile phone needs, Reanud and van Biljon (2008) conclude that older mobile phone users are not a homogenous group, and therefore their worth requirements cannot be addressed by reducing functionality blindly. In this sense of limiting functionality without thinking harder, Fuchsberger et al. (2012) in a study of intergenerational computer-mediated communication argue that simplicity does not mean to have a very reduced interface, but a logical one. If we examine technology use by older people over extended periods of time, previous research has revealed that social and digital inclusion, and independence (i.e. conducting tasks without support) are important motivations for *many* active older people to use and explore "ordinary" technologies (Sayago et al., 2011). Once they learn to use digital technologies, it is not uncommon to see most older people exploring (see Fig. 3.2) what else they can do with digital technologies (Sayago et al., 2011).

Fig. 3.2 Older adult with years of experience of using computers—and teaching other older adults how to use computers—is exploring a new tablet while waiting for a drop-in session to begin. Photo by the author

Simplicity is a valuable design principle—for all users. However, when designing for older adults, there is room to revisit and critically reflect on what simplicity truly means.

As Norman (2016) argues, the simplicity or complexity of a technology is not determined by the number of features it offers, but by whether users have a clear conceptual model of how it works. Many older adults, like other users, do not wish to sacrifice the power, flexibility, and even complexity of digital technologies. Instead of reducing functionality, Norman proposes an alternative: taming complexity. This approach acknowledges that complexity is an inherent and necessary part of life, and that thoughtful design can make complex systems usable and meaningful.

In line with this perspective, this book advocates for a deeper understanding of older adults and their relationships with digital technologies. Rather than defaulting to simplicity, designers could aim to create systems that are intuitive, empowering, and adaptable, supporting older adults in engaging with technology on their own terms.

3.3.3 Special Needs

Another widespread assumption in technology design for older adults is that they have different or unique needs. This belief is reflected in research that frames older adults as having distinct requirements compared to the general population (Batbold et al., 2024; Tang et al., 2025). It is also common to encounter the notion that older adults have "special" needs (Jiang et al., 2024; Rebola & Jones, 2013). But what exactly are these needs? Where do they originate? And what makes them "special" or "different"?

Changing capabilities associated with ageing are often cited as the basis for older adults' so-called special or different needs. Age-related declines in functional abilities can lead many older individuals to require support in maintaining independence. These challenges are compounded by environmental barriers, such as the general inaccessibility of urban spaces (Righi et al., 2015; Van Hoof et al., 2018), and by the design of digital technologies and online services, which have historically overlooked older adults (Newell & Gregor, 2000).

Some of the so-called special needs of older adults may not stem solely from ageing-related changes but may also be shaped—or even created—by specific technologies. For example, Sayago et al. (2016) argue that many older adults are uninterested in video games that involve violent or fantasy scenarios, such as killing aliens, because these games do not align with their values or interests. This suggests that some older adults may have distinct game-playing preferences, which are often overlooked in mainstream game design.

Other needs arise from the marginalization that many older adults experience, such as undesired isolation. Barbosa Neves et al. (2019) highlight how loneliness and social isolation are deeply relational and multidimensional phenomena, shaped by both structural and personal factors. Additionally, the lived experience of ageing—something younger people do not share—can bring forward values such as purpose, belonging, competence, contribution, and independence (Leong & Robertson, 2016). These values often inform older adults' expectations and interactions with technology and should be considered in design processes that aim to be inclusive and meaningful.

However, are there truly significant distinctions between the needs of older adults and those of the rest of the population?

On one hand, older adults are often perceived as having different or "special" needs due to their marginalization. As discussed in Chap. 2, older people are frequently viewed as belonging to a separate world—one that many of us do not engage with until we reach later life. This distancing can lead to the perception that older adults' needs are fundamentally different. Moreover, and from a historical perspective (Hazan, 1994), a significant part of most of our meaningful encounters with older adults tend to occur in contexts that confront the boundary between life and death, reinforcing a sense of otherness.

On the other hand, many needs commonly attributed to older adults, such as the desire for belonging, independence, and competence, are shared across all age groups. The need for technical support, for example, is not exclusive to ageing-related performance deficits. It can also arise from unfamiliarity, situational impairments, or poor design (Yu & Chattopadhyay, 2024). Research on mobile news applications shows that usability, navigation, visual presentation, readability, and interaction design affect not only older adults but most users (Jiang et al., 2024).

The medicalization of ageing, which frames older adults as being in constant need of help, reinforces the notion of "special" needs. Yet, special needs exist across all demographics. Teenagers, middle-aged adults, children, and people with disabilities all have unique needs shaped by their interactions with digital technologies. Their needs are not inherently negative, burdensome, or "different", they are simply contextual and emergent.

3.3.4 Solutionism

Technology is widely regarded as a solution to the perceived problems or "special needs" of many older adults (Bradwell et al., 2024; Gerling et al., 2020; Vines et al., 2015). As Neven and Peine (2017) discuss, ageing is frequently positioned as a societal crisis—one that demands intervention through innovation. Investing in technological solutions is often framed as a strategic decision that yields benefits across multiple levels, from economic growth to healthcare efficiency. This ageing-and-innovation discourse also shapes policy on active and independent ageing, influencing both research agendas and system design.

However, this narrative has been increasingly challenged. Scholars such as Brophy (2015) argue that digital technologies should not only address problems but also enrich and add meaning to the lives of older adults. Neven and Peine (2017) further emphasize the need to critically examine how technologies position older people and whether they genuinely serve their interests.

This section aligns with these concerns and contributes to the discussion from a slightly different perspective: should all problems associated with ageing be "fixed"?

Morozov (2013) reminds us that "not everything that could be fixed should be fixed", arguing that "the opportunity to err, to sin, to do the wrong thing: all of these are constitutive of human freedom, and any concentrated attempt to root them out will root out that freedom as well". As discussed in Chap. 2, ageing is fundamentally about change; it is a dynamic process involving both losses and gains. If technology is positioned solely as a means to solve older adults' problems, we risk reducing ageing to a set of deficits to be corrected. This raises a critical question: what remains of the experience of ageing and older age if all its challenges are "fixed"?

This is not to suggest that technology should not be designed to support older adults. However, it is important to consider the unintended, and perhaps unexpected, consequences of what Morozov (2013) describes as the "mindset that recasts every instance of an efficiency deficit as an obstacle that needs to be overcome". For example, in a feasibility study of a novel communication technology aimed at enhancing social connectedness among older adults in residential care (Barbosa-Neves et al., 2019), most outcomes were positive. Yet, and as discussed by Barbara and colleagues, the study also revealed that the app augmented feelings of inadequacy and heightened awareness of frailty for at least two participants. Additionally, tensions with relatives regarding differing communication preferences and expectations emerged. These kinds of negative outcomes are rarely addressed in research, despite their relevance to understanding the full impact of technological interventions.

3.3.5 Critical Corner

Widespread assumptions about older adults and technology can, and should, be challenged. Many older people may not require simpler technologies, despite what common sense or prevailing narratives suggest. Likewise, the needs of older adults may not be as "special" or "different" as often portrayed. Designing technologies to solve problems is both justified and necessary: older adults face low levels of accessibility across many aspects of life, and age-related changes in functional abilities can create situations where support is needed. In these cases, technology can play a vital role in improving quality of life.

However, digital technologies should not be limited to fixing problems. They should also aim to enrich the lives and values of older adults (Knowles & Hanson, 2018a, 2018b; Knowles et al., 2021; Lazar et al., 2016; Liu et al., 2016). Increasingly, older adults themselves are calling for technologies that support meaningful engagement, personal growth, and social connection, as reflected in the literature. This shift invites designers and researchers to move beyond deficit-based models and towards more inclusive, empowering, and value-driven approaches to technology design.

The emphasis on special needs aligns with the broader social discourse on ageing, as discussed in Chap. 2. The assumption of simplicity can also be understood through the lens of problematizing older adults. They are often portrayed as requiring simple, and distinct, technologies because they are perceived as "other" users, whose needs are presumed to be known and predefined. In certain contexts, framing an issue as a problem can be constructive. For instance, a user interface that is difficult to navigate constitutes a problem for its users; simplifying it is a logical solution. However, framing older adults themselves as a problem is, at best, troubling. It is disrespectful and potentially harmful, reinforcing stereotypes and undermining their agency.

Human–Computer Interaction (HCI) has made significant strides in recent years. On one hand, there is a growing call within the field for reflective awareness regarding contexts in which computational technologies may be inappropriate or even harmful (Baumer & Silberman, 2011). Older adults are increasingly being conceptualized as human and social actors, rather than merely as a collection of human factors. This shift aligns with contemporary HCI design approaches that extend beyond problem-solving to embrace more nuanced understandings of users. As the ageing population continues to grow, interactions with older adults are likely to become more frequent, potentially leading to more accurate, and hopefully, more humane, representations of ageing and older individuals.

On the other hand, many researchers and designers remain outsiders to the lived experiences of older adults (Carmichael et al., 2007; Petrie, 2023; Rice et al., 2007). To avoid assumptions and oversimplifications in technology design, a deeper, more holistic understanding of older age and the ageing process is—and will be—essential.

3.4 Ways of Designing

The literature identifies four primary approaches to designing technologies for older adults. This chapter categorizes them as follows: deficit-driven design, technology generation and familiarity design, ageing experience design, and participatory design.

3.4.1 Deficit-Driven Design

Deficit-driven, or deficit-focused, design (Carroll et al., 2012; Gerling et al., 2020) concentrates on compensating for age-related declines in functional abilities (see Fig. 3.3). Within this framework, older adults are often treated primarily as human factors to be accommodated.

Fig. 3.3 Older adult is struggling to highlight a sentence by using the mouse. His hands are shaky. Keyboard-based interaction works better for him. Photo by the author

Design guidelines are one of the most illustrative examples of this approach. Fisk et al. (2009), which specifically address sensory, cognitive, perceptual, and mobility changes associated with ageing, put forward guidelines that explore the implications of such changes for the design of both products and environments. Other seminal contributions include the work of Chadwick-Dias et al. (2007), who developed a set of empirically grounded web design guidelines based on five years of research involving over 200 older adults. These guidelines aim to support the design of web interfaces tailored to the needs of older users, recommending practices such as scalable text, high-contrast visuals, clearly identifiable links, and simplified terminology. Money et al. (2009) focus on adapting online forms to promote inclusive access for older adults. Their design recommendations, such as enlarging click-sensitive areas around radio buttons and reducing the need for scrolling, address specific age-related changes in functional abilities. Similarly, Hanson et al. (2006) explore software adaptations that accommodate predictable age-related challenges in vision, hearing, motor skills, and cognition. Examples of these adaptations include enlarging content and interface controls to improve usability for older users.

With a strong emphasis on web accessibility, the *WAI Guidelines and Older Web Users* compare design recommendations for making web pages usable by older adults—based on age-related changes in capabilities—with the Web Content Accessibility Guidelines (WCAG) developed by the Web Accessibility Initiative (WAI).[3] These recommendations, derived from a comprehensive literature review, are organized under four key principles: perceivable, operable, understandable, and robust. In both of these foundational references, older adults are primarily conceptualized as a set of human factors to be accommodated.

Design guidelines and recommendations are receiving growing research attention, which extends beyond web technologies to other domains, including digital television (Nunes et al., 2012), touchscreens and mobile devices (Caird et al., 2008; Haesner et al.,

---

[3] https://www.w3.org/WAI/WAI-AGE/comparative.html.

2018), mobile apps (Gomez-Hernandez et al., 2023), and health applications (Liu et al., 2021). There are also ongoing efforts to integrate existing usability and design guidelines intended for older adults into a single collection (Lindberg & De Troyer, 2021).

As discussed by Gerling et al. (2020), a deficit-focused perspective is also evident in numerous technology solutions aimed at supporting physical activity among older adults, as published in prominent HCI venues such as TACCESS, UIST, CHI, and TOCHI. These systems typically address physical activity within the context of health, often framing older users in terms of their functional limitations.

3.4.2 Technology Generation and Familiarity Design

This design approach emphasizes technology generation (Ivan et al., 2020) and familiarity with prior technologies (Ellis & Marshall, 2019). It recognizes that many older adults have experience with technologies from earlier generations, such as mechanical or electromechanical systems, and that the knowledge and skills acquired during their formative years (Docampo et al., 2001) often do not transfer easily, if at all, to modern digital technologies. By focusing on older adults' previous technological experiences,[4] this approach highlights the challenges they face when learning to use contemporary systems.

As discussed by Docampo et al. (2001), older adults' difficulties in using contemporary technologies stem primarily from three factors: user interface complexity, age-related effects, and technology generation. The first two have been addressed previously. The third technology generation refers to the evolution of interaction styles over time, from physical interfaces such as push buttons and switches to software-based interfaces involving displays, touchscreens, and gesture- or voice-based controls, where only a portion of the functionality is immediately visible.

Technology generation is shaped during a person's formative years, typically between the ages of 10 and 25, through repeated interactions with technological interfaces. These early experiences tend to influence technology use later in life (Sackmann & Winkler, 2013). Once a skill or pattern of interaction is learned, it can be difficult to unlearn or adapt to new paradigms[5] (Lim, 2010). Several studies provide evidence that technology generation significantly affects how older adults engage with digital technologies. For instance, behaviours such as trial-and-error, which is common among users familiar with

---

[4] Technology generation does not exist in a vacuum. Contextual aspects, such as level of education and country, shape it, and could be considered to develop a more nuanced view of this concept (Ivan et al., 2020).

[5] Yet, and even though different age groups cannot evade their historical past, it does not mean that they are restricted to it. In other words, belonging to a previous technology generation does not mean that one cannot actively participate in contemporary technology generations (Costa et al., 2019).

software-based systems, are rarely observed in individuals from electromechanical generations, who may avoid experimentation due to the perceived consequences of pressing the wrong button (Docampo et al., 2001).

Familiarity, which is rooted in both experience and understanding, and it is often shaped by generational exposure to technology (Ellis & Marshall, 2019), is another key factor contributing to older adults' difficulties with digital technologies. While younger generations grow up surrounded by computers, smartphones, and digital devices, older adults—particularly those who retired before the digital revolution—may lack exposure to these technologies (Turner & Walle, 2006).

Research has shown that familiarity with earlier technologies influences design preferences (Frohlich et al., 2016; Lee et al., 2024), supports the development of interfaces for late-life online communication (Brewer et al., 2016), facilitates technology adoption (Suopajärvi, 2015), reduces anxiety when engaging with new technologies (Rodríguez et al., 2019), and helps older users orient themselves within digital applications (Hollinworth & Hwang, 2011a, 2011b).

3.4.3 Ageing Experience Design

This design approach, which has gained increasing attention over time, moves beyond deficit-driven and technology generation perspectives to embrace a more holistic understanding of ageing (Yang & Moffatt, 2024). For instance, *design for situated elderliness* (Subasi et al., 2013) proposes defining ageing not by biological age or institutional categories, but through the lens of everyday practices and lived experiences (see Fig. 3.4).

Brewer and Piper (2016), in their study of online content creation, advocate for *designing for late-life development*, framing older adulthood as a phase of growth, reflection, and continued engagement. Similarly, Romero et al. (2010) emphasize the importance of

Fig. 3.4 Same older adult who experienced difficulties using the mouse is watching a video of a TV series that he could not watch the night before. Photo by the author

designing for transitions when developing persuasive technologies for older adults living with frailty. Their approach highlights the need to account for the nuanced and dynamic nature of older adults' lives and their social networks.

Central to this approach is the idea that design should not be limited to supporting functional independence but should also enhance the experience of living (Brophy, 2015). For example, Strengers et al., (2022) emphasize the importance of designing for older adults' enjoyment, curiosity, pleasure, and play with smart devices, viewing these as pathways to developing digital living skills and promoting positive ageing.

Gaver et al. (2011) similarly argue that design can support *ludic experiences*, such as engagement, sociability, and enjoyment, rather than focusing solely on physical care needs, particularly in care home settings. Light et al. (2016) further contend that design should not introduce new challenges but instead enable individuals to continue doing what they have learned to value.

Within this design approach, Nicenboim et al. (2018) view older adults as resourceful individuals and introduce the *connected resources* approach. This perspective shifts the focus from solving problems to supporting strategies, and from designing for specific scenarios to enabling a variety of uses. It stands in contrast to the deficit-driven model, which often frames ageing as a problem and older adults as frail or passive.

While technological solutions based on compensation and prevention may offer value in certain contexts, they are unlikely to remain appropriate over time. Designs that assume older adults require "foolproof" systems risk reinforcing limiting stereotypes. In response, *designing for resourceful ageing* seeks to continuously reinvent the practical arrangements of everyday life—not only in response to physical changes, but also in alignment with evolving preferences and desires (Kitazaki et al., 2019).

To counter the tendency to design technologies that may be unacceptable or stigmatizing for older adults, Knowles et al. (2021) advocate for designing *for the experience of ageing*—that is, for the transitions, changes, and lived experiences that occur across the lifespan. Caldeira et al. (2022) similarly argue that aligning technology with older adults' current coping strategies is key to supporting ageing in place, especially as individuals move through different stages of ageing: active, adaptive, and passive.

Ferri et al. (2017) also emphasize ageing as a process, calling for design approaches that raise awareness, challenge ageist stereotypes, and address the felt experience of ageing. Expanding on this perspective, Willatt et al. (2024) propose an *intersectional life course lens* to deepen understanding of older adults' histories, lived experiences, and socio-material contexts. This lens supports the development of design methods and technologies that offer meaningful value to older adults' lives.

3.4.4 Involving Older Adults in Design

"The inclusion rather than exclusion of older adults in the design process and research of digital technology is essential if technology is to fulfil the promise of improving well-being" (Mannheim et al., 2019, p. 1). From co-design workshops and co-creation studies to participatory design approaches, a great deal of research involves older adults in technology design activities, e.g. (Ambe et al., 2019a, 2019b; Antony et al., 2023; Davidson & Jensen, 2013; Hakobyan et al., 2015; Ostrowski et al., 2024; Sheahan et al., 2024; Smith, 2025; Weber et al., 2023). Thus, and inspired by the seminal work of Vines et al. (2013), it is important to examine how extant research involves older people in technology design.

More than 15 years ago, Massimi et al., (2007) discussed a conflict in participatory design with older adults. This conflict, which involves three viewpoints on designing for older people, is still relevant today, as it can potentially affect how we recruit older adults, run participatory or co-design sessions with them, and the results and/or technology being designed (e.g. "we are developing a system targeted at people aged 60–75 vs. a system for the community you belong to"):

- "Design for me", e.g. we are developing a system targeted at people like you - this suggests personalized technologies which are difficult to be used by others.
- "Design for us", e.g. we are developing a system for the community you belong to— this suggests inclusive technologies.
- "Design for them", e.g. we are developing a system for people of your age that are homebound – this promotes generalization (older people as a whole) and the idea that the participants do not form part of this general group.

Mannheim et al. (2019) argue that "inclusion does not necessarily mean that research is conducted free of ageism". Specifically, it is not enough to co-design. It is also about how it is done. The setting in which the study takes place (Cajamarca et al., 2022) as well as the stage in which older people are involved, and the presence or lack of facilitators (Pramudito et al., 2023), play an important role in the outcome (see Fig. 3.5).

Frennert and Östlund (2014) examine how older adults are involved in the design of social robots, finding that they are implicated but not present in the development of these technologies. The results also show that the conception of older people is plagued with stereotypical views such as that they are lonely, frail and in need of robotic assistance. Ageist and interventionist perspectives on the imagined reality of being older seem to dictate the primary use case of social assistive robots for care and alleviation of loneliness. Older adults as users are positioned as passive in this imagined HRI (Mannheim et al., 2025).

A systematic review of involving older users in technology design (Fischer et al., 2020) reveals that there is "an overwhelming inclination to involve older people as informants, testers and consultants". The consequences of their involvement are learning,

Fig. 3.5 Group consisting of older adults, an undergraduate student and the author, participating in the conceptualisation of a prototype of an e-mail system. The session was conducted in the bar of the social centre where the older participants took computer sessions. The bar was a place where they got-together before and after the sessions. Thanks to the author's involvement in the social centre, where he conducted ethnographic research for 4 years, the participants felt empowered and made interesting contributions to the design of the prototype: an e-mail system for older (and not so old) people. Photo by the author

adjusted design and an increased sense of participation, and not either acceptance or adoption. These results call into question the widely held assumption that user involvement inevitably yields beneficial outcomes. There might be unexpected consequences too. For instance, participating can evoke feelings related to disappointment in terms of (1) not fully living up to the expectations older adults have of themselves, (2) being afraid of disappointing project team members, and (3) not gaining the kind of competences expected as a result of their participation (Viklund et al., 2023).

3.4.5 Critical Corner

A range of design approaches exists for developing technologies for older adults. Some are primarily problem-oriented, aligning with traditional human-centred design principles and reflecting the historical roots of HCI in fields such as human factors, ergonomics, and usability engineering. These approaches often reinforce solutionism, prioritizing the resolution of perceived deficits. In contrast, more contemporary HCI approaches, such as asset-based (Wong-Villacres et al., 2021) and value-based design (Friedman & Hendry, 2019), focus on strengths, values, and lived experiences. Other emerging perspectives, like more-than-human design and post-growth HCI, remain mostly underexplored in the context of ageing.

Each design approach tends to conceptualize older adults differently, with distinct objectives and assumptions that often appear disconnected from one another. For instance, when designing for fun, curiosity, or empowerment, age-related changes in capabilities may play a secondary role. This diversity in design perspectives illustrates the complexity of designing digital technologies for older adults; it is far from a straightforward task.

This complexity also raises important questions: Should design for ageing prioritize functionality, lived experience, or both? And how can we ensure that technologies reflect the evolving identities, preferences, and needs of older adults without reinforcing stereotypes or stigmas?

It is tempting to consider whether these diverse design approaches could, or should, be integrated into a unified framework to simplify the development of digital technologies for older adults. After all, changing capabilities, technology generation, familiarity, and ageing as a process are all integral to older adults' experiences with technology. However, whether such integration would reduce design complexity or represent the most effective path forward remains an open question, one that is highly context-dependent and warrants further research.

What does seem clear is that designing technologies for older adults cannot rely solely on any single perspective—be it changing capabilities, generational familiarity, or the lived experience of ageing. Crucially, the involvement of older adults in the design process is essential. Without their participation, even well-intentioned designs risk reinforcing stereotypes or failing to meet real needs.

In addition to the diversity of design approaches, further research is needed to explore their unintended or unexpected consequences. For example, do approaches focused on compensating for changing capabilities and technology generation inadvertently reinforce stereotypes, such as the notion that older adults inherently need help, or that they are more interested in the past than the future? Similarly, do approaches that emphasize positive or active ageing risk marginalizing those in the so-called "fourth age" by focusing too heavily on the "third age" ideals of fun, curiosity, pleasure, and empowerment?

Critical questions also arise around the influence of social policies that promote active and independent ageing, which may unintentionally reinforce deficit-focused perspectives in technology design. How can we navigate or challenge these policy-driven narratives to create more inclusive and nuanced technologies for older adults?

Rather than discouraging inquiry, these questions highlight the richness and complexity of designing for ageing. They underscore that this is not only a challenging design space but also a fertile and evolving area of research, one that invites continued exploration, reflection, and innovation.

Importantly, involving older adults in technology design processes, while widely advocated, is not sufficient on its own to meaningfully shape outcomes. Their participation often remains superficial, typically limited to an informant role. As Östlund et al. (2015)

argue, a key challenge lies in whether we are prepared to address deeper issues of social change and power relations within design. Without this shift, older people risk being positioned within constructed hopes for technological solutions, some celebrated as innovations, others dismissed as failures.

This calls for a more critical and inclusive approach to participation, one that moves beyond tokenism and engages older adults (and members of their local communities) (Righi et al., 2017) as co-creators with agency and expertise. Only by doing so can we begin to challenge dominant narratives and foster technologies that genuinely reflect the diverse realities of ageing.

3.5   Ways of Understanding Technology

In older-adult centred design, two distinct ways of understanding technology can be identified.

A strand of research conceptualizes digital technologies as *technologies for doing, i.e.* tools primarily designed to assist or enable older adults in performing specific tasks or activities. These are often instrumental in nature, focusing on functionality and support. When interaction with older adults is framed through the lens of human factors, digital technologies are typically designed as *technologies for doing*. These technologies support older adults in engaging with computer-enabled activities such as taking and sharing photos online (Apted et al., 2006; Coelho et al., 2017; Cornejo et al., 2013) or sending emails (Dickinson et al., 2005). To facilitate these interactions, it is essential to compensate for age-related changes in functional abilities through thoughtful user interface design. Consequently, design approaches focused on changing capabilities are particularly prominent in this context. Assistive technologies, which aim to support older adults in performing daily living tasks, are clear examples of *technologies for doing*.[6] These designs prioritize usability and accessibility, often emphasizing functionality over emotional or experiential aspects.

Another strand of research extends the focus of digital technologies beyond task completion and assistance (Simão et al., 2024) to support goals related to *being* and *feeling*, such as empowerment (Duarte De Camargo et al., 2024; Lazar et al., 2016; Rogers & Marsden, 2013), enjoyment (Strengers et al., 2022), imagining sustainable futures (Lengyel et al., 2023), and fostering agency (Fitzpatrick et al., 2015). While age-related changes in functional abilities still play a role in achieving these goals, other factors, such as experiential values of technologies (Desai et al., 2022), personal motivations, life course trajectories, and the stage of ageing, are equally important. The design

---

[6] Whilst assistive technologies might belong to the "rhetoric of compassion" (Rogers & Marsden, 2013), the view of technology for doing is not another funny word for it; the examples above show technologies designed for older people who are not conceptualised as in need of help.

approaches discussed in Sect. 3.3, which focus on the experiences of ageing, are particularly well-suited to enabling *technologies for being and feeling*. These approaches move beyond instrumental functionality to support older adults in living meaningful, emotionally rich, and socially connected lives.

3.5.1 Critical Corner

Older adults, ageing, and technology are mutually constitutive, i.e. they shape and define one another (Peine & Neven, 2021). Shifting conceptualizations and representations of older adults, alongside evolving technology design approaches, influence the meanings and purposes of digital technologies in this field.

As stated in (Verbeek, 2015), humans and technologies should not be viewed as separate entities between which interaction occurs; rather, they are co-produced through interaction. As discussed in this chapter, technologies for older adults are shaped by prevailing ideas about ageing, such as those centred on changing capabilities or life stages, and, conversely, these technologies also shape societal views of ageing, including notions of positive ageing and distinctions between the third and fourth age. What is being designed is not merely a technological artefact, but a human-world relation.

We become what we build for ourselves (Frauenberger, 2020). In older-adult centred design, we are constructing multiple relationships between people, technologies, and the social and environmental worlds they inhabit. Future research could explore the extent to which these relationships align with the aspirations and values of the next generation of older adults, especially given that much existing research continues to focus on today's ageing population, and their impact on advancing towards more sustainable futures.

References

Abeele,V. V.,Schraepen, B.,Huygelier, H.,Gillebert, C.,Gerling, K., & VanEe, R. (2021). Immersive virtual reality for older adults: Empirically grounded design guidelines. *ACM Transactions on Accessible Computing, 14*(3), 1–30. https://doi.org/10.1145/3470743

Álvarez-García, J., Durán-Sánchez, A., Del Río-Rama, M. D. L. C., & Correa-Quezada, R. (2019). Older adults and digital society: Scientific coverage. *International Journal of Environmental Research and Public Health, 16*(11), 2010. https://doi.org/10.3390/ijerph16112010

Ambe, A. H., Brereton, M., Soro, A., Buys, L., & Roe, P. (2019a). The adventures of older authors: Exploring futures through co-design fictions. In *Proceedings of the 2019 CHI conference on human factors in computing systems* (pp. 1–16). https://doi.org/10.1145/3290605.3300588

Ambe, A. H., Brereton, M., Soro, A., Chai, M. Z., Buys, L., & Roe, P. (2019b). Older people inventing their personal internet of things with the IoT Un-Kit experience. In *Proceedings of the 2019 CHI conference on human factors in computing systems* (pp. 1–15). https://doi.org/10.1145/3290605.3300552

Antony, V. N., Cho, S. M., & Huang, C.-M. (2023). Co-designing with older adults, for older adults: robots to promote physical activity. In *Proceedings of the 2023 ACM/IEEE international conference on human-robot interaction* (pp. 506–515). https://doi.org/10.1145/3568162.3576995

Apted, T., Kay, J., & Quigley, A. (2006). Tabletop sharing of digital photographs for the elderly. In *Proceedings of the SIGCHI conference on human factors in computing systems* (pp. 781–790). https://doi.org/10.1145/1124772.1124887

Ashby, S., Hanna, J., Matos, S., Nash, C., & Faria, A. (2019). Fourth-Wave HCI Meets the 21st Century Manifesto. *Proceedings of the Halfway to the Future Symposium 2019,* 1–11. https://doi.org/10.1145/3363384.3363467

Ashcroft, A., & Knowles, B. (2025). Beyond binary: Re-imagining age using gender in HCI for inclusive design. In *Proceedings of the extended abstracts of the CHI conference on human factors in computing system*s (pp. 1–7). https://doi.org/10.1145/3706599.3719766

Bannon, L. J. (1991). From human factors to human actors the role of psychology and human-computer interaction studies in systems design. In J. Greenbaum & M. Kyng (Eds.), *Design at work: Cooperative design of computer systems* (pp. 25–44). Lawrence Erlbaum Associates.

Barbosa Neves, B., Franz, R., Judges, R., Beermann, C., & Baecker, R. (2019). Can digital technology enhance social connectedness among older adults? A feasibility study. *Journal of Applied Gerontology, 38*(1), 49–72. https://doi.org/10.1177/0733464817741369

Batbold, T., Soro, A., & Schroeter, R. (2024). Mentorable interfaces for automated vehicles: A new paradigm for designing learnable technology for older adults. In *Proceedings of the CHI conference on human factors in computing systems* (pp. 1–15). https://doi.org/10.1145/3613904.3642390

Baumer, E. P. S., & Brubaker, J. R. (2017). Post-userism. In *Proceedings of the 2017 CHI conference on human factors in computing systems* (pp. 6291–6303). https://doi.org/10.1145/3025453.3025740

Baumer, E. P. S., & Silberman, M. S. (2011). When the implication is not to design (technology). In *Proceedings of the SIGCHI conference on human factors in computing systems* (pp. 2271–2274). https://doi.org/10.1145/1978942.1979275

Bødker, S., Dindler, C., Iversen, O. S., & Smith, R. C. (2022). Participatory design. *Springer International Publishing.* https://doi.org/10.1007/978-3-031-02235-7

Bødker, (2015). Third wave HCI, 10 years later. *Interactions, 22,* 5 (September-October 2015), 24–31. https://doi.org/10.1145/2804405

Bradwell, H. L., Cooper, L., Baxter, R., Tomaz, S., Edwards, K. J., Whittaker, A. C., et al. (2024). Implementation of virtual reality motivated physical activity via omnidirectional treadmill in a supported living facility for older adults: A mixed- methods evaluation.: virtual reality to motivate physical activity for older adults. In *Proceedings of the CHI conference on human factors in computing systems* (pp. 1–13). https://doi.org/10.1145/3613904.3642281

Brewer, R., & Piper, A. M. (2016). 'Tell it like it really is': A case of online content creation and sharing among older adult bloggers. In *Proceedings of the 2016 CHI conference on human factors in computing systems* (pp. 5529–5542). https://doi.org/10.1145/2858036.2858379

Brewer, R., Garcia, R. C., Schwaba, T., Gergle, D., & Piper, A. M. (2016). Exploring traditional phones as an e-mail interface for older adults. *ACM Transactions on Accessible Computing, 8*(2), 1–20. https://doi.org/10.1145/2839303

Brophy, C. (2015). *Aging and everyday technology.*

Burema, D. (2022). A critical analysis of the representations of older adults in the field of human–robot interaction. *AI & SOCIETY, 37*(2), 455–465. https://doi.org/10.1007/s00146-021-01205-0

Caird, J. K., Chisholm, S. L., & Lockhart, J. (2008). Do in-vehicle advanced signs enhance older and younger drivers' intersection performance? Driving simulation and eye movement results. *International Journal of Human-Computer Studies, 66*(3), 132–144. https://doi.org/10.1016/j.ijhcs.2006.07.006

Cajamarca, G., Herskovic, V., Lucero, A., & Aldunate, A. (2022). A co-design approach to explore health data representation for older adults in chile and Ecuador. In *Designing interactive systems conference* (pp. 1802–1817). https://doi.org/10.1145/3532106.3533558

Caldeira, C., Nurain, N., & Connelly, K. (2022). "I hope I never need one": Unpacking stigma in aging in place technology. In *CHI conference on human factors in computing systems* (pp. 1–12). https://doi.org/10.1145/3491102.3517586

Carmichael, A., Newell, A. F., & Morgan, M. (2007). The efficacy of narrative video for raising awareness in ICT designers about older users' requirements. *Interacting with Computers, 19*(5–6), 587–596. https://doi.org/10.1016/j.intcom.2007.06.001

Carroll, J. M., Convertino, G., Farooq, U., & Rosson, M. B. (2012). The firekeepers: Aging considered as a resource. *Universal Access in the Information Society, 11*(1), 7–15. https://doi.org/10.1007/s10209-011-0229-9

Chadwick-Dias, A., Bergel, M., & Tullis, T. S. (2007). Senior Surfers 2.0: A reexamination of the older web user and the dynamic web. In C. Stephanidis (Ed.), *Universal access in human computer interaction. Coping with diversity* (Vol. 4554, pp. 868–876). Springer Berlin Heidelberg. https://doi.org/10.1007/978-3-540-73279- 2_97

Chin, J., & Fu, W.-T. (2012). Age differences in exploratory learning from a health information website. In *Proceedings of the SIGCHI conference on human factors in computing systems* (pp. 3031–3040). https://doi.org/10.1145/2207676.2208715

Chin, J., Fu, W.-T., & Kannampallil, T. (2009). *Adaptive information search: Age-dependent interactions between cognitive profiles and strategies.*

Cila, N., Smit, I., Giaccardi, E., & Kröse, B. (2017). Products as agents: Metaphors for designing the products of the IoT age. In *Proceedings of the 2017 CHI conference on human factors in computing systems* (pp. 448–459). https://doi.org/10.1145/3025453.3025797

Coelho, J., Rito, F., & Duarte, C. (2017). "You, me & TV"—Fighting social isolation of older adults with Facebook, TV and multimodality. *International Journal of Human-Computer Studies, 98*, 38–50. https://doi.org/10.1016/j.ijhcs.2016.09.015

Cornejo, R., Favela, J., & Tentori, M. (2010). Ambient displays for integrating older adults into social networking sites. In G. Kolfschoten, T. Herrmann, & S. Lukosch (Eds.), *Collaboration and technology* (Vol. 6257, pp. 321–336). Springer Berlin Heidelberg. https://doi.org/10.1007/978-3-642-15714-1_24

Cornejo, R., Tentori, M., & Favela, J. (2013). Enriching in-person encounters through social media: A study on family connectedness for the elderly. *International Journal of Human-Computer Studies, 71*(9), 889–899. https://doi.org/10.1016/j.ijhcs.2013.04.001

Costa, C., Gilliland, G., & McWatt, J. (2019). 'I want to keep up with the younger generation'—older adults and the web: A generational divide or generational collide? *International Journal of Lifelong Education, 38*(5), 566–578. https://doi.org/10.1080/02601370.2019.1678689

Davidson, J. L., & Jensen, C. (2013). *Participatory design with older adults: An analysis of creativity in the design of mobile healthcare applications.* C&C.

Desai, S., McGrath, C., McNeil, H., Sveistrup, H., McMurray, J., & Astell, A. (2022). Experiential value of technologies: A qualitative study with older adults. *International Journal of Environmental Research and Public Health, 19*(4), 2235. https://doi.org/10.3390/ijerph19042235

Dickinson, A., Newell, A., Smith, M., & Hill, A. (2005). Introducing the Internet to the over-60s: Developing an email system for older novice computer users. *Interacting with Computers, 17*(6), 621–642.

Docampo Rama, M., Ridder, H. D., & Bouma, H. (2001). Technology generation and age in using layered user interfaces. *Gerontechnology, 1*(1), 25–40. https://doi.org/10.4017/gt.2001.01. 01.003.00

Du, Q., Song, Z., Jiang, H., Wei, X., Weng, D., & Fan, M. (2024). LightSword: A customized virtual reality exergame for long-term cognitive inhibition training in older adults. In *Proceedings of the CHI conference on human factors in computing systems* (pp. 1–17). https://doi.org/10.1145/361 3904.3642187

Duarte De Camargo, J., Silva, T., & Ferraz De Abreu, J. (2024). Empowering older adults: A user-centered approach combining iTV and voice assistants to promote social interactions. In A. Marcus, E. Rosenzweig, & M. M. Soares (Eds.), *Design, user experience, and usability* (Vol. 14714, pp. 13–25). Springer Nature Switzerland. https://doi.org/10.1007/978-3-031-61356-2_2

Dunee, A., & Raby, F. (2013). Speculative everything. In *Design, fiction, and social dreaming*. The MIT Press.

Durick, J., Robertson, T., Brereton, M., Vetere, F., & Nansen, B. (2013). Dispelling ageing myths in technology design. In *Proceedings of the 25th Australian computer-human interaction conference: Augmentation, application, innovation, collaboration* (pp. 467–476). https://doi.org/10.1145/254 1016.2541040

Ellis, A., & Marshall, M. T. (2019). Can skeuomorphic design provide a better online banking user experience for older adults? *Multimodal Technologies and Interaction, 3*(3), 63. https://doi.org/ 10.3390/mti3030063

Eriksson, E., Yoo, D., Bekker, T., & Nilsson, E. M. (2024). More-than-human perspectives in human-computer interaction research: A scoping review. In *Nordic conference on human-computer interaction* (pp. 1–18). https://doi.org/10.1145/3679318.3685408

Ferri, G., Bardzell, J., & Bardzell, S. (2017). Rethinking age in HCI through anti-ageist playful interactions. *Interacting with Computers, 29*(6), 779–793. https://doi.org/10.1093/iwc/iwx012

Fischer, B., Peine, A., & Östlund, B. (2020). The importance of user involvement: A systematic review of involving older users in technology design. *The Gerontologist, 60*(7), e513–e523. https://doi.org/10.1093/geront/gnz163

Fisk, A., Rogers, W., Charness, N., Czaja, S., & Sharit, J. (2009). Designig for older adults. In *Principles and creative human factors approaches* (2nd ed.). CRS Press.

Fitzpatrick, G., Huldtgren, A., Malmborg, L., Harley, D., & Ijsselsteijn, W. (2015). Design for agency, adaptivity and reciprocity: reimagining AAL and telecare agendas. In V. Wulf, K. Schmidt, & D. Randall (Eds.), *Designing socially embedded technologies in the real-world* (pp. 305–338). Springer London. https://doi.org/10.1007/978-1-4471-6720-4_13

Frauenberger, C. (2020). Entanglement HCI the next wave? *ACM Transactions on Computer-Human Interaction, 27*(1), 1–27. https://doi.org/10.1145/3364998

Frennert, S., & Östlund, B. (2014). Review: Seven matters of concern of social robots and older people. *International Journal of Social Robotics, 6*(2), 299–310. https://doi.org/10.1007/s12369-013-0225-8

Friedman, B., & Hendry, D. (2019). *Value sensitive design*. MIT Press.

Frohlich, D. M., Lim, C., & Ahmed, A. (2016). Co-designing a diversity of social media products with and for older people. In *Proceedings of the 7th international conference on software development and technologies for enhancing accessibility and fighting info-exclusion* (pp. 323–330).

Fuchsberger, V., Sellner, W., Moser, C., & Tscheligi, M. (2012). Benefits and hurdles for older adults in intergenerational online interactions. In K. Miesenberger, A. Karshmer, P. Penaz, & W. Zagler (Eds.), *Computers helping people with special needs* (Vol. 7382, pp. 697–704). Springer Berlin Heidelberg. https://doi.org/10.1007/978-3-642-31522-0_104

Gaver, W., Boucher, A., Bowers, J., Blythe, M., Jarvis, N., Cameron, D., et al. (2011). The photo-stroller: Supporting diverse care home residents in engaging with the world. In *Proceedings of the SIGCHI conference on human factors in computing systems* (pp. 1757–1766). https://doi.org/10.1145/1978942.1979198

Gerling, K., Ray, M., Abeele, V. V., & Evans, A. B. (2020). Critical reflections on technology to support physical activity among older adults: An exploration of leading HCI venues. *ACM Transactions on Accessible Computing, 13*(1), 1–23. https://doi.org/10.1145/3374660

Ghoshal, S., & Dasgupta, S. (2023). Design values in action: Toward a theory of value dilution. In *Proceedings of the 2023 ACM designing interactive systems* conference (pp. 2347–2361). https://doi.org/10.1145/3563657.3596122

Giaccardi, E., Redström, J., & Nicenboim, I. (2025). The making(s) of more- than-human design: Introduction to the special issue on more-than-human design and HCI. *Human-Computer Interaction, 40*(1–4), 1–16. https://doi.org/10.1080/07370024.2024.2353357

Gomez-Hernandez, M., Ferre, X., Moral, C., & Villalba-Mora, E. (2023). Design guidelines of mobile apps for older adults: Systematic review and thematic analysis. *JMIR mHealth and uHealth, 11*, Article e43186. https://doi.org/10.2196/43186

Guo, P. J. (2017). Older adults learning computer programming: Motivations, frustrations, and design opportunities. In *Proceedings of the 2017 CHI conference on human factors in computing systems* (pp. 7070–7083). https://doi.org/10.1145/3025453.3025945

Haesner, M., Wolf, S., Steinert, A., & Steinhagen-Thiessen, E. (2018). Touch interaction with google glass—Is it suitable for older adults? *International Journal of Human-Computer Studies, 110*, 12–20. https://doi.org/10.1016/j.ijhcs.2017.09.006

Hakobyan, L., Lumsden, J., & O'Sullivan, D. (2015). Participatory design: How to engage older adults in participatory design activities. *International Journal of Mobile Human Computer Interaction, 7*(3), 78–92. https://doi.org/10.4018/ijmhci.2015070106

Hanson, V. L., Snow-Weaver, A., & Trewin, S. (2006). Software personalization to meet the needs of older adults. *Gerontechnology, 5*(3), 160–169. https://doi.org/10.4017/gt.2006.05.03.005.00

Harrison, S., Tatar, D., & Sengers, P. (n.d.). *The three paradigms of HCI.*

Hassenzahl, M. (2010). *Experience design: Technology for all the right reasons.* Springer.

Hawthorn, D. (2000). Possible implications of aging for interface designers. *Interacting with Computers, 12*, 507–528.

Hazan, H. (1994). *Old age constructions and deconstructions.* Cambridge University Press.

Hollinworth, N., & Hwang, F. (2011a). Cursor relocation techniques to help older adults find 'lost' cursors. In *Proceedings of the SIGCHI conference on human factors in computing systems* (pp. 863–866). https://doi.org/10.1145/1978942.1979068

Hollinworth, N., & Hwang, F. (2011b). Investigating familiar interactions to help older adults learn computer applications more easily. In *Proceedings of HCI 2011 The 25th BCS Conference on Human Computer Interaction.* https://doi.org/10.14236/ewic/HCI2011.79

Hummert, M. L. (2011). Age stereotypes and aging. In *Handbook of the psychology of aging* (pp. 249–262). Elsevier. https://doi.org/10.1016/B978-0-12-380882-0.00016-4

Ivan, L., Loos, E., & Bird, I. (2020). The impact of 'technology generations' on older adults' media use: Review of previous empirical research and a seven-country comparison. *Gerontechnology, 19*(4), 1–19. https://doi.org/10.4017/gt.2020.19.04.387

Jiang, Q., Deng, L., Zhang, J., & Pengbo, Y. (2024). User-centered design strategies for age-friendly mobile news apps. *SAGE Open, 14*(4), 21582440241285390. https://doi.org/10.1177/21582440241285393

Kitazaki, M., Nicenboim, J., & Giaccardi, E. (2019). Connected resources—Empowering older people to age resourcefully. In *Extended Abstracts of the 2019 CHI conference on human factors in computing systems (CHI EA '19). Association for Computing Machinery*, New York, NY, USA, Paper VS06, 1. https://doi.org/10.1145/3290607.3311774

Knowles, B., & Hanson, V. L. (2018a). The wisdom of older technology (non)users. *Communications of the ACM, 61*(3), 72–77. https://doi.org/10.1145/3179995

Knowles, B., & Hanson, V. L. (2018b). Older adults' deployment of 'distrust.' *ACM Transactions on Computer-Human Interaction, 25*(4), 1–25. https://doi.org/10.1145/3196490

Knowles, B., Hanson, V. L., Rogers, Y., Piper, A. M., Waycott, J., Davies, N., Ambe, A. H., Brewer, R. N., Chattopadhyay, D., Dee, M., Frohlich, D., Gutierrez- Lopez, M., Jelen, B., Lazar, A., Nielek, R., Pena, B. B., Roper, A., Schlager, M., Schulte, B., & Yuan, I. Y. (2021). The harm in conflating aging with accessibility. *Communications of the ACM, 64*(7), 66–71. https://doi.org/10.1145/3431280

Lazar, A., Brewer, R. N., & Knowles, B. (2025). HCI and older adults: The critical turn and what comes next. *Foundations and Trends® in Human-Computer Interaction, 19*(2), 112–212. https://doi.org/10.1561/1100000094

Lazar, A., Cornejo, R., Edasis, C., & Piper, A. M. (2016). Designing for the third hand: Empowering older adults with cognitive impairment through creating and sharing. In *Proceedings of the 2016 ACM conference on designing interactive systems* (pp. 1047–1058). https://doi.org/10.1145/2901790.2901854

Lee, C. H., Lee, J., Kim, D., Kim, I., & Song, H. (2024). Designing self-ordering kiosk for older adults: Familiarity design focusing on representation, manipulation, and organization. *Computers in Human Behavior, 156*, Article 108236. https://doi.org/10.1016/j.chb.2024.108236

Lee, H. R., Cheon, E., Lim, C., & Fischer, K. (2022). Configuring humans: What roles humans play in HRI research. In *2022 17th ACM/IEEE international conference on human-robot interaction (HRI)* (pp. 478–492). https://doi.org/10.1109/HRI53351.2022.9889496

Lengyel, D., Kaipainen, K., Sturdee, M., Heron, M., Lewis, M., & Liddle, J. (2023). Hands-on workshop on tabletop role-playing for inclusive design: Imagining sustainable futures for 'older adults'. In *26th international academic mindtrek conference* (pp. 289–293). https://doi.org/10.1145/3616961.3616977

Leong, T. W., & Robertson, T. (2016). Voicing values: Laying foundations for ageing people to participate in design. In *Proceedings of the 14th participatory design conference: Full papers* (Vol. 1, pp. 31–40). https://doi.org/10.1145/2940299.2940301

Li, S., Blythe, P., Zhang, Y., Edwards, S., Xing, J., Guo, W., Ji, Y., Goodman, P., & Namdeo, A. (2021). Should older people be considered a homogeneous group when interacting with level 3 automated vehicles? *Transportation Research Part f: Traffic Psychology and Behaviour, 78*, 446–465. https://doi.org/10.1016/j.trf.2021.03.004

Light, A., Pedell, S., Robertson, T., Waycott, J., Bell, J., Durick, J., & Leong, T. W. (2016). What's special about aging. *Interactions, 23*(2), 66–69. https://doi.org/10.1145/2886011

Lim, C. S. C. (2010). Designing inclusive ICT products for older users: Taking into account the technology generation effect. *Journal of Engineering Design, 21*(2–3), 189–206. https://doi.org/10.1080/09544820903317001

Lindberg, R. S. N., & De Troyer, O. (2021). Towards an up to date list of design guidelines for elderly users. In *CHI Greece 2021: 1st International Conference of the ACM Greek SIGCHI Chapter, 1–7*. https://doi.org/10.1145/3489410.3489418

Liu, N., Purao, S., & Tan, H.-P. (2016). Value-inspired service design in elderly home-monitoring systems. *IEEE International Conference on Pervasive Computing and Communication Workshops (PerCom Workshops), 2016*, 1–6. https://doi.org/10.1109/PERCOMW.2016.7457138

Liu, N., Yin, J., Tan, S.S.-L., Ngiam, K. Y., & Teo, H. H. (2021). Mobile health applications for older adults: A systematic review of interface and persuasive feature design. *Journal of the American Medical Informatics Association, 28*(11), 2483–2501. https://doi.org/10.1093/jamia/ocab151

Mannheim, I., Edan, Y., Alon, R., Peine, A., & Nimrod, G. (2025). Ageism, power, and gaps: Co-designing robots with older adults. *International Journal of Social Robotics.* https://doi.org/10.1007/s12369-025-01325-3

Mannheim, I., Schwartz, E., Xi, W., Buttigieg, S. C., McDonnell-Naughton, M., Wouters, E. J. M., & Van Zaalen, Y. (2019). Inclusion of older adults in the research and design of digital technology. *International Journal of Environmental Research and Public Health, 16*(19), 3718. https://doi.org/10.3390/ijerph16193718

Maskeliūnas, R., Damaševičius, R., & Segal, S. (2019). A review of internet of things technologies for ambient assisted living environments. *Future Internet, 11*(12), 259. https://doi.org/10.3390/fi11120259

Massimi, M., Baecker, R. M., & Wu, M. (2007). Using participatory activities with seniors to critique, build, and evaluate mobile phones. In: *Proceedings of the 9th international ACM SIGACCESS conference on computers and accessibility* (pp. 155–162). https://doi.org/10.1145/1296843.1296871

McLaughlin, A., & Pak, R. (2020). *Designing displays for older adults (Second edition).* CRC Press.

Mitzner, T. L., Boron, J. B., Fausset, C. B., Adams, A. E., Charness, N., Czaja, S. J., Dijkstra, K., Fisk, A. D., Rogers, W. A., & Sharit, J. (2010). Older adults talk technology: Technology usage and attitudes. *Computers in Human Behavior, 26*(6), 1710–1721. https://doi.org/10.1016/j.chb.2010.06.020

Moffatt, K., & McGrenere, J. (2010). Steadied-bubbles: Combining techniques to address pen-based pointing errors for younger and older adults. In *Proceedings of the SIGCHI conference on human factors in computing systems* (pp. 1125–1134). https://doi.org/10.1145/1753326.175349

Money, A. G., Fernando, S., Lines, L., & Elliman, A. D. (2009). Developing and evaluating web-based assistive technologies for older adults. *Gerontechnology, 8*(3), 165–177. https://doi.org/10.4017/gt.2009.08.03.013.00

Mori, K., & Harada, E. T. (2010). Is learning a family matter?: Experimental study of the influence of social environment on learning by older adults in the use of mobile phones. *Japanese Psychological Research, 52*(3), 244–255. https://doi.org/10.1111/j.1468-5884.2010.00434.x

Morozov, E. (2013). *To save everything, click here. Public Affairs.*

Neven, L. (2010). 'But obviously not for me': Robots, laboratories and the defiant identity of elder test users. *Sociology of Health & Illness, 32*(2), 335–347. https://doi.org/10.1111/j.1467-9566.2009.01218.x

Neven, L., & Peine, A. (2017). From triple win to triple sin: How a problematic future discourse is shaping the way people age with technology. *Societies, 7*(3), 26. https://doi.org/10.3390/soc7030026

Newell, A. F., & Gregor, P. (2000). "User sensitive inclusive design"—In search of a new paradigm. In *Proceedings on the 2000 conference on universal usability - CUU '00* (pp. 39–44). https://doi.org/10.1145/355460.355470

Ng, R., & Indran, N. (2022). Not too old for tiktok: How older adults are reframing aging. *The Gerontologist, 62*(8), 1207–1216. https://doi.org/10.1093/geront/gnac055

Ng, R., Allore, H. G., Trentalange, M., Monin, J. K., & Levy, B. R. (2015). Increasing negativity of age stereotypes across 200 years: Evidence from a database of 400 million Words. *PLoS ONE, 10*(2), Article e0117086. https://doi.org/10.1371/journal.pone.0117086

Nicenboim, I., Kitazaki, M., Kihara, T., Torralba Marin, A., & Havranek, M. (2018). Connected resources: A novel approach in designing technologies for older people. In *Extended abstracts of the 2018 CHI conference on human factors in computing systems* (pp. 1–4). https://doi.org/10.1145/3170427.3186527

Norman, D (2023). *Design for a better world.*

Norman, D. (2005). *Human-centered design considered harmful. Interactions.*

Norman, D. (2016). *Living with complexity*. The MIT Press.

Nunes, F., Kerwin, M., & Silva, P. A. (2012). *Design recommendations for tv user interfaces for older adults: Findings from the eCAALYX project*.

Östlund, B., Olander, E., Jonsson, O., & Frennert, S. (2015). STS-inspired design to meet the challenges of modern aging. Welfare technology as a tool to promote user driven innovations or another way to keep older users hostage? *Technological Forecasting and Social Change, 93*, 82–90. https://doi.org/10.1016/j.techfore.2014.04.012

Ostrowski, A. K., Zhang, J., Breazeal, C., & Park, H. W. (2024). Promising directions for human-robot interactions defined by older adults. *Frontiers in Robotics and AI, 11*, 1289414. https://doi.org/10.3389/frobt.2024.1289414

Pang, C., Collin Wang, Z., McGrenere, J., Leung, R., Dai, J., & Mo@att, K. (2021). Technology adoption and learning preferences for older adults: Evolving perceptions, ongoing challenges, and emerging design opportunities. In *Proceedings of the 2021 CHI conference on human factors in computing systems* (pp. 1–13). https://doi.org/10.1145/3411764.3445702

Park, D., & Schwarz, N. (2000). *Cognitive aging*. Taylor & Francis.

Peine, A., & Neven, L. (2021). The co-constitution of ageing and technology—A model and agenda. *Ageing and Society, 41*(12), 2845–2866. https://doi.org/10.1017/S0144686X20000641

Petrie, H. (2023). Talking 'bout my Generation... or not?: The digital technology life experiences of older people. In *Extended Abstracts of the 2023 CHI conference on human factors in computing systems* (pp. 1–9). https://doi.org/10.1145/3544549.3582742

Pramudito, A., Barbareschi, G., & Sato, C. (2023). Enhancing self-reflection in older adults through collage making. In *Proceedings of the international conference on the internet of things* (pp. 236–239). https://doi.org/10.1145/3627050.3631568

Rebola, C. B., & Jones, B. (2013). Sympathetic devices: Designing technologies for older adults. In *Proceedings of the 31st ACM international conference on design of communication* (pp. 151–156). https://doi.org/10.1145/2507065.2507083

Renaud, K., & van Biljon, J. (2008). *Predicting technology acceptance and adoption by the elderly: A qualitative study*.

Rice, M., Newell, A., & Morgan, M. (2007). Forum Theatre as a requirement gathering methodology in the design of a home telecommunication system for older adults. *Behaviour & Information Technology, 26*(4), 323–331. https://doi.org/10.1080/01449290601177045

Righi, V., Sayago, S., & Blat, J. (2015). Urban ageing: Technology, agency and community in smarter cities for older people. In *Proceedings of the 7th international conference on communities and technologies (C&T '15). Association for computing machinery, New York, NY, USA* (pp. 119–128). https://doi.org/10.1145/2768545.2768552

Righi, V., Sayago, S., & Blat, J. (2017). When we talk about older people in HCI, who are we talking about? Towards a 'turn to community' in the design of technologies for a growing ageing population. *International Journal of Human-Computer Studies, 108*, 15–31.

Rodríguez, I., Karyda, M., Lucero, A., & Herskovic, V. (2019). Aestimo: A tangible kit to evaluate older adults' user experience. In D. Lamas, F. Loizides, L. Nacke, H. Petrie, M. Winckler, & P. Zaphiris (Eds.), *Human-Computer Interaction–INTERACT 2019* (Vol. 11746, pp. 13–32). Springer International Publishing. https://doi.org/10.1007/978-3-030-29381-9_2

Rogers, Y., & Marsden, G. (2013). Does he take sugar?: Moving beyond the rhetoric of compassion. *Interactions, 20*(4), 48–57. https://doi.org/10.1145/2486227.2486238

Rogers, Y., Paay, J., Brereton, M., Vaisutis, K. L., Marsden, G., & Vetere, F. (2014). Never too old: Engaging retired people inventing the future with MaKey MaKey. In *Proceedings of the SIGCHI conference on human factors in computing systems* (pp. 3913–3922). https://doi.org/10.1145/2556288.2557184

Romero, N., Sturm, J., Bekker, T., De Valk, L., & Kruitwagen, S. (2010). Playful persuasion to support older adults' social and physical activities. *Interacting with Computers, 22*(6), 485–495. https://doi.org/10.1016/j.intcom.2010.08.006

Sackmann, R., & Winkler, O. (2013). Technology generations revisited: The internet generation. *Gerontechnology, 11*(4), 493–503. https://doi.org/10.4017/gt.2013.11.4.002.00

Sayago, S., Forbes, P., & Blat, J. (2013). Older people becoming successful ICT learners over time: Challenges and strategies through an ethnographical lens. *Educational Gerontology, 39*(7), 527–544.

Sayago, S. (2019). *Perspectives on human-computer interaction research with older people.* Springer Cham.

Sayago, S., & Bergantiños, A. (2021). Exploring the first experiences of computer programming of older people with low levels of formal education: A participant observational case study. *International Journal of Human-Computer Studies, 148*, Article Number: 102577. https://doi.org/10.1016/j.ijhcs.2020.102577

Sayago, S., Rosales, A., Righi, V., Ferreira, S., Coleman, G., & Blat, J. (2016). On the conceptualization, design and evaluation of appealing, meaningful and playable digital games for older people. *Games and Culture, 11*(1–2), 53–80.

Sayago, S., Sloan, D., & Blat, J. (2011). Everyday use of computer-mediated communication tools and its evolution over time: An ethnographical study with older people. *Interacting with Computers, 23*(5), 543–554.

Sharma, V., Kumar, N., & Nardi, B. (2024). Post-growth human–computer interaction. *ACM Translated Computer-Human Interaction, 31*(1), Article 9 (November 2023), 37. https://doi.org/10.1145/3624981

Sharit, J., Hernández, M. A., Czaja, S. J., & Pirolli, P. (2008). Investigating the roles of knowledge and cognitive abilities in older adult information seeking on the web. *ACM Transactions on Computer-Human Interaction, 15*(1), 1–25. https://doi.org/10.1145/1352782.1352785

Sheahan, J., Cardamone, E., & Vines, J. (2024). Exploring generative postcard futures with older adults. In *Adjunct Proceedings of the 2024 nordic conference on human-computer interaction* (pp. 1–6). https://doi.org/10.1145/3677045.3685502

Shneiderman, B. (2022). *Human-centered AI.* Oxford University Press.

Simão, H., Bernardino, A., Guerreiro, T., & Forlizzi, J. (2024). Beyond assistance: Robots aligned with older adults' values. In *Companion of the 2024 ACM/IEEE international conference on human-robot interaction* (pp. 145–147). https://doi.org/10.1145/3610978.3638357

Smith, K. (2025). *Touching experiences: How older adults envision ambient and tangible social technology through the lens of time.* CHI, Yokohama Japan.

Soden, R., Toombs, A., & Thomas, M. (2024). Evaluating interpretive research in HCI. *Interactions, 31*(1), 38–42. https://doi.org/10.1145/3633200

Strengers, Y., Duque, M., Mortimer, M., Pink, S., Martin, R., Nicholls, L., et al. (2022). "Isn't this Marvelous": Supporting older adults' wellbeing with smart home devices through curiosity, play and experimentation. In *Designing interactive systems conference* (pp. 707–725). https://doi.org/10.1145/3532106.3533502

Subasi, Ö., Fitzpatrick, G., Malmborg, L., & Östlund, B. (2013). Design culture for ageing well: Designing for 'situated elderliness'. In A. Holzinger, M. Ziefle, M. Hitz, & M. Debevc (Eds.), *Human Factors in Computing and Informatics* (Vol. 7946, pp. 581–584). Springer Berlin Heidelberg. https://doi.org/10.1007/978-3-642-39062- 3_36

Suhaimi, N. M., Zhang, Y., Joseph, M., Kim, M., Parker, A. G., & Gri@in, J. (2022). Investigating older adults' attitudes towards crisis informatics tools: Opportunities for enhancing community resilience during disasters. In *CHI Conference on human factors in computing systems* (pp. 1–16). https://doi.org/10.1145/3491102.3517528

Suopajärvi, T. (2015). Past experiences, current practices and future design. *Technological Forecasting and Social Change, 93*, 112–123. https://doi.org/10.1016/j.techfore.2014.04.006

Tang, E., Song, T., Zhu, Z., Li, J., & Lee, Y.-C. (2025). *AI literacy education for older adults: Motivations, challenges and preferences.* CHI EA 25.

Tang, X., Ding, X. (Sharon), & Zhou, Z. (2023). Towards equitable online participation: A case of older adult content creators' role transition on short-form video sharing platforms. In *Proceedings of the ACM on human-computer interaction* (Vol. 7(CSCW2), pp. 1–22). https://doi.org/10.1145/3610216

Tanprasert, T., Dai, J., & McGrenere, J. (2024). HelpCall: Designing informal technology assistance for older adults via videoconferencing. In *Proceedings of the CHI conference on human factors in computing systems* (pp. 1–23). https://doi.org/10.1145/3613904.3642938

Turner, P., & Walle, G. V. D. (2006). Familiarity as a basis for universal design. *Gerontechnology, 5*(3), 150–159. https://doi.org/10.4017/gt.2006.05.03.004.00

Van Hoof, J., Kazak, J. K., Perek-Białas, J. M., & Peek, S. T. M. (2018). The challenges of urban ageing: Making cities age-friendly in Europe. *International Journal of Environmental Research and Public Health, 15*(11), 2473. https://doi.org/10.3390/ijerph15112473

Verbeek, P.-P. (2015). *Beyond interaction: A short introduction to mediation theory.*

Viklund, E. W. E., Nilsson, I., Hägglund, S., Nyholm, L., & Forsman, A. K. (2023). The perks and struggles of participatory approaches: Exploring older persons' experiences of participating in designing and developing an application. *Gerontechnology, 22*(1), 1–12. https://doi.org/10.4017/gt.2023.22.1.816.03

Vines, J., Clarke, R., Wright, P., McCarthy, J., & Olivier, P. (2013). Configuring participation: On how we involve people in design. In *Proceedings of the SIGCHI conference on human factors in computing systems* (pp. 429–438). https://doi.org/10.1145/2470654.2470716

Vines, J., Pritchard, G., Wright, P., Olivier, P., & Brittain, K. (2015). An age-old problem: Examining the discourses of ageing in HCI and strategies for future research. *ACM Transactions on Computer-Human Interaction, 22*(1). https://doi.org/10.1145/2696867

Wakkary, R. (2021). *Things we could design for more than human-centered worlds.* The MIT Press.

Weber, P., Mahmood, F., Ahmadi, M., Von Jan, V., Ludwig, T., & Wieching, R. (2023). Fridolin: Participatory design and evaluation of a nutrition chatbot for older adults. *I-Com, 22*(1), 33–51. https://doi.org/10.1515/icom-2022-0042

While, Z., Blascheck, T., Gong, Y., Isenberg, P., & Sarvghad, A. (2024). Glanceable data visualizations for older adults: Establishing Thresholds and examining disparities between age groups. In *Proceedings of the CHI conference on human factors in computing systems* (pp. 1–17). https://doi.org/10.1145/3613904.3642776

Wiberg, M., & Teigland, R. (2024). Computing for the 22nd Century: More-than-human to see the environmental footprints of pro- found technologies. In *Halfway to the Future (HTTF '24), October 21–23, 2024, Santa Cruz, CA, USA. ACM, New York, NY, USA* (p. 4). https://doi.org/10.1145/3686169.3686177

Wiles, J. L., & Jayasinha, R. (2013). Care for place: The contributions older people make to their communities. *Journal of Aging Studies, 27*(2), 93–101. https://doi.org/10.1016/j.jaging.2012.12.001

Willatt, A., Gray, S. I., Manchester, H., Foster, T., & Cater, K. (2024). An intersectional lifecourse lens and participatory methods as the foundations for co-designing with and for minoritised older adults. *Proceedings of the ACM on Human-Computer Interaction, 8*(CSCW1), 1–29. https://doi.org/10.1145/3637295

Wong-Villacres, M., Gautam, A., Tatar, D., & DiSalvo, B. (2021). Reflections on assets-based design: A journey towards a collective of assets-based thinkers. *Procceding of ACM Human-Computer Interaction, 5*(401), 32. https://doi.org/10.1145/347954

Wu, Z., Wang, D., Zhang, S., Huang, Y., Wang, Z., & Fan, M. (2024). Toward making virtual reality (VR) more inclusive for older adults: Investigating aging E@ect on target selection and manipulation tasks in VR. In *Proceedings of the CHI conference on human factors in computing systems* (pp. 1–17). https://doi.org/10.1145/3613904.3642558

Xu, W., Dainoff, M. J., Ge, L., & Gao, Z. (2023). Transitioning to human interaction with AI systems: New challenges and opportunities for HCI professionals to enable human-centered AI. *International Journal of Human-Computer Interaction, 39*(3), 494–518. https://doi.org/10.1080/10447318.2022.2041900

Yang, M., & Mo@att, K. (2024). Navigating the maze of routine disruption: Exploring how older adults living alone navigate barriers to establishing and maintaining physical activity habits. In *Proceedings of the CHI conference on human factors in computing systems* (pp. 1–15). https://doi.org/10.1145/3613904.3642842

Yu, J. E., & Chattopadhyay, D. (2024). Reducing the search space on demand helps older adults find mobile UI features quickly, on par with younger adults. In *Proceedings of the CHI conference on human factors in computing systems* (pp. 1–22). https://doi.org/10.1145/3613904.3642796

Zhao, W., Kelly, R. M., Rogerson, M. J., & Waycott, J. (2024). Older adults imagining future technologies in participatory design workshops: Supporting continuity in the pursuit of meaningful activities. In *Proceedings of the CHI conference on human factors in computing systems* (pp. 1–18). https://doi.org/10.1145/3613904.3641887

Methodological Configurations 4

Abstract

This chapter builds upon the interpretative overview of older-adult centred design presented in the previous chapter by examining methodological configurations that shape our work with older adults. It critically engages with the HCI literature addressing methodological challenges in research involving older participants and reflects on the limited progress and attention this topic has received in recent years, despite its relevance for fostering a culture of transparency and enhancing methodological rigour

Keywords

Research methods • Evaluation • Engagement • Requirements • Focus groups • Participatory design

4.1 Purpose and Context

Much of our understanding of older adults' interactions with digital technologies in HCI stems from the application of design and research methods, as well as their participation and engagement in these processes. Consequently, the methodological issues involved in conducting research and design activities with older adults warrant careful consideration. This chapter offers an interpretative overview of these issues, with the aim of encouraging HCI researchers to adopt a more critical and transparent approach to methodological configurations in both their work with older adults and their scholarly publications. To situate this chapter into a broader context, the following section provides a brief overview of methodological configurations in both HCI and ageing.

4.1.1 Methodological Configurations in HCI

Methodological configurations and reflections are not rare in HCI. As Lazar et al. (2017) note, HCI has a long-standing tradition of borrowing, adapting, and developing research methods from various disciplines to establish what is considered acceptable practice within the field. For example, the HCI literature has consistently emphasized the need to tailor research and design methods when working with children (Giannakos et al., 2022; Lehnert et al., 2022). Engaging people with disabilities in HCI research also requires careful attention to methodological configurations and ethical considerations (Lazar et al., 2017). Aligning with several recent calls (Da Silva Junior et al., 2024; Oppenlaender et al., 2025) for reflecting on the ways in which HCI research is conducted and reported, Qi and Yu (2025) reveal misconceptions of how Participatory Design has been carried out in HCI thus far, such as overlooking the political commitment to democracy and empowerment, and narrowly equating PD to design ideation or prototyping. Lee et al. (2022) reflect on how humans are configured in Human–Robot Interaction (HRI) research, finding that they have mostly played passive roles and suggesting opportunities for reflexivity, e.g. socioeconomically underserved populations are one of the groups whose voices and knowledge are largely overlooked. Meissner (2025) reflects on participatory research with low-resources communities and argues for re-imagining community-based participatory research as an opportunity for fostering give-and-take relationships with participants. The vulnerability of researchers (Boldi et al., 2023) and reflections on ethical encounters in HCI (Petrie, 2025; Waycott et al., 2015), with increasing research being conducted in sensitive settings, are also receiving increasing research attention in the field. Drawing inspiration from fast and slow thinking, a recent work (Wiberg, 2026) provides a critical examination of the temporalities at play in HCI research approaches and related methodological configurations (e.g. fast and slow research).

4.1.2 Methodological Configurations in Ageing Research

In ageing research (Gubrium & Sankar, 1994; Weil, 2017), there is a broad consensus that research methods and activities involving older adults should be adapted, configured, or conducted in non-standard ways—that is, approaches that diverge from conventional textbook methodologies. An overview of works showing the diversity of methodological configurations is provided next.

Research with older adults of low socioeconomic status faces several methodological challenges, such as difficulties associated with recruitment, high drop-out rates, and literacy problems (Platzer et al., 2021). Older migrants often have this status and (Chen & Buckingham, 2025) discuss methodological challenges in conducting qualitative research with older Chinese migrants in New Zealand, ranging from recruitment, which relied heavily on trust-building facilitated by gatekeepers, and the research implementation, in which collaborative questionnaires and dialogical interviews enabled the participants to co-construct the research procedure. These challenges concur with the methodological and ethical issues discussed in (Bilecen & Fokkema, 2022), which divides them into five groups: defining the population of interest, sampling, selection of questions, validity of the answers and composition of the research team. Lörinc et al. (2022) also report on lessons learned from conducting walking interviews with older migrants in England, including the issue of anonymity and ethics-in-practice, because of the unpredictable nature of walking interviews.

Age-related changes in vision have a strong impact on eye tracking studies. Shoenfelt et al. (2025) discuss challenges and best practices for conducting eye tracking with older adults, covering pre-screening and participant selection as well as experimental design, data processing and analysis. Shoenfelt et al. (2025) recommend age-tailored experimental eye tracking studies. Vision loss also affects narrative engagement, and Mathiesen et al. (2025) highlight the need adapt research methodologies and measurement scales to suit older adults and auditory narratives, ensuring they capture unique aspects of auditory engagement and account for sensory impairments. Trujillo et al. (2018) also argue that to accommodate older visually impaired adults in research activities, researchers must, in general, be sensitive to them and find alternatives to obtaining written consent, use good diction and cueing to interview them.

There are also methodological challenges in physical activity research with older adults, from sampling issues, including fear of exacerbation of prior medical conditions and poor perceived benefits, to data measures and collection, such as unfamiliarity with technologies like pedometers and a lack of motivation for maintaining activity logs (Chase, 2013).

With respect to methods, studies that conduct research methods gaining popularity in qualitative and participatory research, such as photovoice (Ferlatte et al., 2022), with older adults, report on methodological reflections, such as the need to adjust the size of discussion groups and build and maintain trust throughout the study. Quinn (2010) shows that surveys of adults over the age of 55 have unique methodological considerations, which typically concern the physiological and psychological factors associated with age-related declines in cognitive functioning and health. Remote usability testing methods need to be adapted too when conducted with older people, on the grounds of changing capabilities and experience with digital technologies (Hill et al., 2021a, 2021b).

Cultural issues, which have been found to determine the ways in which research methods in HCI and other areas is conducted (Sayago, 2023), are also expected to come into play in research involving older participants. For example, Mehta (2011) shows that one of the most important difficulties in running focus groups with older Asian adults is that many Asian people are inhibited about sharing personal problems in a group context.

4.1.3 Methodological Configurations in Older-Adult HCI

Drawing on the previous two sections, there are reasons to believe that conducting HCI research with older adults demands a thoughtful approach to selecting and adapting methods that facilitate respectful interaction and the collection of high-quality data. This claim was made two decades ago (Eisma et al., 2004). However, the methodologies we employ for working with older people in our research activities in HCI are largely the same as those used for any other user group. This does not seem to be in accord with the discourse on special needs (see Chapter 3) and the body of knowledge accumulated in design (e.g. guidelines) for building technologies tailored to older adults. Also, it is rare to see a discussion on adaptations or configurations needed to conduct research methods with participants who deviate from the typical user (a young or middle-aged, abled individual from WEIRD countries) in leading HCI textbooks,[1] such as *The Handbook of Usability Testing* (Rubin, 1994), *Interaction design: Beyond Human–Computer Interaction* (Sharp et al., 2023), *Human–Computer Interaction* (Rogers et al., 1994), *The Human–Computer Interaction Handbook* (Jacko, 2012), *Modern Statistical Methods for HCI* (Robertson & Kaptein, 2016), *Contextual Design: Design for Life* (Holtzblatt & Beyer, 2017) and *Ways of Knowing in HCI* (Olson & Kellog, 2014). A noteworthy exemption is *Research Methods in Human–Computer Interaction* (Lazar et al., 2017), which features a chapter about working with research participants with disabilities, which is to be commended, but it has no chapter devoted to older adults.

4.2 Working with Older Research Participants in HCI: An Overview

The HCI literature focused on methodological issues involved in working with older research participants[2] can be divided into: (i) requirements elicitation and focus groups, (ii) involvement or engagement of older people in research activities, (iii) technology evaluation, (iv) research logistics, and (v) participatory design.

4.2.1 Requirements Elicitation and Focus Groups

The studies outlined in this section focus on requirements elicitation, which is a critical activity in the development of digital technologies. Most of the studies employed focus groups for requirements gathering. All these studies point out that eliciting requirements from older adults is challenging and that adaptations or configurations in requirements

---

[1] In other research textbooks, such as the *Routledge International Handbook of Participatory Design* (Simonsen & Robertson, 2013), *Researching Social Life* (Gilbert, 2008), *Analyzing Social Settings. A Guide to Qualitative observations and Analysis* (Lofland et al., 2006), *The Handbook of Interview Research* (Gubrium & Holstein, 2001), *The Oxford and SAGE Handbook of Qualitative Research* (Nathan, 2014), it is more common to witness an appreciation towards diversity among research participants (and researchers too).

[2] Given that the focus on methodological challenges and difficulties might easily veer—unwillingly—into ageism, or might potentially give that impression, this section aims to discuss these previous works in a non-ageist way.

elicitation methods, such as questionnaires, interviews, and focus groups, are needed. Central to these adaptations are age-related changes in functional abilities and intergenerational communication, along with other significant issues related to older people's participation in research activities, such as motivations, anxiety, and expectations.

By drawing on the lessons learned from developing effective methods for the early involvement of older people in the development of digital technologies for people aged 60 and over, Eisma et al. (2004) argue that eliciting requirements for products which do not yet exist from older people poses unique challenges. These were often due to their previous experience of digital technology use. Changing capabilities due to ageing and self-reporting issues (Park & Schwarz, 2000) came also into play. Eisma et al. (2004) also note that motivations behind older adults' participation in research activities should be considered. Older research participants might take part in research activities for socializing (see Fig. 4.1). Thus, researchers should be aware of this fact and adjust the activity accordingly so that it is useful for research purposes too. The authors also found that the best way of addressing reluctance while filling out questionnaires was for a researcher to administer the questionnaire directly.

Administering questionnaires face-to-face was found to augment their value in another study too (Goodman, 2003). For focus groups to be effective, one of the key issues was to make the participants aware of their expertise, and of how valuable their contribution was to the project. In-home interviews were also found to be an excellent means of discovering

Fig. 4.1 Author having lunch with the lovely participants of a course. This get-together was especially important for the older participants. In addition to learning about computers and digital technologies, the participants attended the lessons and visited the centre to socialize. It was important for the author to be aware of this fact to conduct research activities in a way that were meaningful for both the participants and the research study

information, especially with those older adults who live in frailty, as they likely spend more time in their homes. These lessons learned point out that" existing methodologies are only partly appropriate (…). Traditional methods of user-centred design need to be adapted if they are to enable researchers to effectively elicit requirements from older people" (Goodman, 2003, p. 132).

Methodological issues in running focus groups to elicit user requirements with older adults for an interactive domestic alarm system are also discussed in (Hone & Lines, 2004). The authors argue that running focus groups with a small number (5) of older adults and increased structure—avoiding the use of overly broad, open-ended questions—turned out to be more effective. In the user-centred development of an interactive memory aid, (Inglis et al., 2003) show that focus groups with more than three older people were hard to manage, as age-related changes in functional abilities and difficulties in following the thread of a conversation (because there too many people talking at the same time) proved to hamper many participants' contribution. Still, (Hone & Lines, 2004) argues that focus groups might not be a suitable method for requirements elicitation with older users when the research is highly speculative. As it might happen to all of us, it is difficult be focused "on topic" when little is known about the domain.

Barrett and Kirk (2000) report on the lessons learned from running focus groups with older people with and without disabilities to determine their information needs in terms of maintaining their independence for as long as possible. A number of methodological issues were found, including attendance, duration of discussions, communication and attention problems. The authors recommend that questions be kept simple, and short, and use words that the participants understand. Other relevant recommendations for running focus groups with older people include over-recruitment, allowing for people not showing up on the day, the moderator communicating clearly, listening carefully and sensitively, making participants feel that their responses are valued, and keeping the discussions short so that the older participants are able to sustain attention and interest.

Minocha et al. (2013) discuss the challenges and opportunities of conducting empirical research on the role of online communities on the quality of life and well-being of older people. Minocha and colleagues acknowledge that "we have had to adapt and change our elicitation methods as we have progressed with our empirical research". The older participants liked to be prepared before the interview session, asking ahead of their sessions what questions the researchers were going to ask. They were also anxious about the value of their contributions. In the interviews, the authors noted that their participants preferred conversations where they could relate incidents and stories rather than follow a semi-structured interview protocol.

Dickinson et al. (2003) review methodological issues in gathering accurate requirements with older adults who live in frailty. Traditionally structured focus groups present challenges when the group contains members with sensory or cognitive impairments. Regarding questionnaires, older people are more likely to use 'don't know' options and need a higher threshold of certainty before they will select options useful to the researcher

(Park & Schwarz, 2000). Older people are also less likely to contribute confident opinions about technology, and this may be aggravated by unfamiliar language or settings. To deal with most of these issues, the authors argue that the information gained from observation and discussion in the home is a useful tool for the early stages of requirements gathering, because, among other reasons, the older participants are in a familiar and safe environment, talking to older people when they have the technology close to hand allows them to demonstrate their use of devices to the searcher, which is easier than trying to use words.

4.2.2 Involvement/Engagement

Although gathering requirements with older people can be seen as an example of their involvement or engagement in design processes, the studies outlined in this section (Eisma, 2003; Edlin-White et al., 2012) have a slightly more general focus on how to have older adults involved in the design process of new digital technologies. Recurrent themes are encouraging communication through different activities and flexibility in the methodological approach.

Eisma (2003) introduces the concept of mutual inspiration, which highlights the need to have a common ground between the several parties involved (older people—users, designers and developers) so that dialogue may take place, suggestions are considered and all of them respect the other's contribution and expertise. This common ground is especially important when working with older people, due to age-related changing capabilities, different communication styles and variations in technological experience between them and designers or developers. For example, older adults may be hesitant to criticize or provide negative feedback on products, while developers may struggle to fully understand the impact of age-related impairments.

In Rice et al. (2007), it is discussed a user requirement gathering methodology of provoking discussion among older people by using scenario-based theatre. As stated in Eisma (2003), different styles of communication and knowledge about technology might make it difficult for older people to participate meaningfully in design processes. This methodology enabled older people to engage with the details of a new technological development at pre-prototyping stages in a way that was both comprehensible and enjoyable. It also allowed them to effectively consider and discuss desired functionalities, while bringing up unexpected issues that the researchers had not anticipated. Although originally conceived of as a methodology for requirements gathering, scenario-based drama theatre can also be used in other stages of design processes to increase empathy towards older people, identify areas of improvement in technologies, and keep older people engaged in the project.

Edlin-White et al., (2012) explore appropriate methodologies for involving older people in HCI design processes, in particular, in the MyUI project, which aimed to facilitate the development of highly accessible life-enhancing technology-mediated services for older people. Based on the project experience and literature, some methodological recommendations are made, including "being flexible and open to change while conducting a study, allowing time for off-topic conversations, and defining study measures that are appropriate to older users".

Recently, Marzi et al. (2023) address the question of how older adults prefer to be involved in HCI research. Drawing on an online survey and a series of interviews, the results show that the older research participants were generally willing to participate due to interest in the topic, learning new things and social bonding. However, despite the conceptions within HCI that long-term engagement is crucial within participatory design, older people in the study preferred less commitment and more open formats.

4.2.3 Technology Evaluation

The studies outlined in this section focus mostly on usability evaluation and methods. Recurrent themes are difficulties in conducting widely used usability evaluation methods such as thinking-aloud testing by following standard procedures and the need to adapt methods based on age-related changing abilities, impression management, and real-life use of digital technologies.

Barbosa-Neves et al. (2019) discuss usability and accessibility testing issues that arose in evaluating InTouch, a software application that has a non-language specific interface (based on icons) and supports asynchronous communication, by institutionalized oldest old (80 +) people. The unique characteristics of the user group (older people who live in frailty and are low-digital literate) presented challenges in recruitment. To overcome them, it was instrumental to ensure proxies (staff, caregivers) were engaged since the inception of the study. The thinking-aloud technique proved unsuitable for the participants, as they were too frail to perform tasks while simultaneously verbalizing their thoughts. They also struggled to quantify their responses using the 5-point Likert scale, since this task appeared to be too abstract for them to do. Furthermore, self-presentation and impression management (the need to make a good impression on the researcher) were evident during the testing sessions. Some participants reported that the device was easy to lift, but when asked to perform it they acknowledged that it was more difficult than they had first suggested.

Franz et al. (2018) focus on impression management and its influence on usability studies with older adults, finding that participants performed a range of impression management tactics, such as supplication and exemplification, which call for strategies to manage them and thereby adapt or configure usability testing.

Difficulties in conducting concurrent thinking-aloud were also found during user-centred evaluations of a mobile health system for heart failure self-management with older people (Cornet et al., 2017). The participants spontaneously mentioned difficulty performing the concurrent thinking-aloud technique and explaining their thoughts. For them the

Fig. 4.2 In social centres, individual activities are not so natural as one might think. Individual research methods are more difficult to conduct than methods carried out in pairs or in groups, as these configurations resonate with the natural practices of the older participants in these settings

more frequent researcher-participant exchanges were easier. In fact, the authors point out that these exchanges counteracted the difficulties that the older participants had with the thinking-aloud approach. Franz et al. (2019) compare three variants of the thinking-aloud protocol: concurrent thinking-aloud, retrospective think-aloud and co-discovery with frail older adults, finding that co-discovery is the most suitable think-aloud variant because with it (and only it), older people verbalized their thought process throughout the entire usability test. These results, the authors argue, reinforce the need to understand older people and empower them in usability studies through adjusted methods.

Sayago and Blat (2011) argue that individual research methods, such as questionnaires and standard usability tests are much more difficult to carry out and less useful than social ones to evaluate technologies in real-life settings. Individual interactions are unnatural practices since the older participants of their ethnographic research wanted to be social users (see Fig. 4.2). Filling out questionnaires made the participants feel more isolated and under the impression they were in an examination situation. Interviews or natural exchanges with the researchers were regarded as a much more natural way of establishing social contact and taking part in research activities.

4.2.4   Research Logistics

Research logistics also might need to be adapted when studies involve older people. Franz et al. (2015) note that evaluating technologies with older adults through field studies can present significant logistical challenges, including the enrolment and retention of participants, and access to them due to policies specific to the long-term care homes. Dickinson et al. (2007) provide further considerations for planning and conducting research studies involving older participants, from ensuring directions and instructions for a visit to the research space are clear and explicit to being prepared for participants to ask to bring companions.

4.2.5 Participatory Design, Co-design, and Design Workshops

Participatory design practices have been adapted to address age-related considerations (Vines et al., 2012). A recent scoping review of co-designing conversational agents with older people reveals substantial variability in methods and highlights a lack of consensus on optimal structures for engaging this population (Salomé et al., 2025). Similarly, Regalado et al. (2025) reach comparable conclusions regarding co-creation activities, e.g. "we are lacking a description of the processes and an understanding of how to conduct moments of co-creation with older adults in a continuous and informed way" (p. 144).

Pradhan et al. (2025) report instances of older participants design materials as either patronizing or childish. They also describe cases in which older participants disengaged from designing technologies for themselves, instead focusing on technologies for other older people. This tendency raises important questions about why some older adults may distance themselves from envisioning or co-creating technologies intended for their own use.

Pradhan et al. (2025) argue that when older people do not use design materials provided by the researcher this does not always mean that they stopped engaging with material representations altogether. Instead, participants might have found other ways to express their design ideas, such as using their own bodies as a form of design material. Thus, it might be more important to personalize design materials when conducting design workshops with older people than to rely on standardized or repurposed materials.

Lindsay et al. (2012) report on four challenges when doing participatory design with older adults as an approach to improving the quality of design for this population: maintaining focus and structure in meetings, representing and acting on issues, envisioning tangible concepts, and designing for non-tasks. Massimi et al. (2007) provide another list of considerations for future researchers: provide alternative activities, create temporary subgroups to overcome deficits, minimize crosstalk, make participation an institutional affair, provide activity structure, speed up or down to suit the group, and blend individual and group sessions. Lindsay et al. (2012) reflect on the author's experiences of engaging older adults in the participatory design of smart tools for health information search. Based on their experiences, the authors discuss what worked and what did not, and the approach followed. In terms of recruitment, which is a challenging process (Binda et al., 2018), it was important to enlist allies and work with local communities to recruit older participants instead of recruitment based on flyers posted on and off campus. It was also useful to incorporate design critique (Vines et al., 2012) as an icebreaker and for supporting innovative design ideas. Using a common vocabulary, accommodating schedules and adapting the protocol based on the dynamics of the community centres and participants were also important configurations.

Davidson and Jensen (2013) add to these considerations by aiming to keep design sessions short, allowing for informal socializing, encouraging participation and balancing researcher and participant input. Hakobyan et al. (2015) acknowledge arguments that conducting participatory design with older people can present a number of challenges; finding and recruiting representative participants can prove difficult, conducting observational studies of vulnerable individuals can raise privacy issues, and engaging individuals with impairments can present logistical difficulties. Yet, they argue that it is very difficult to achieve informed and effective design and development of digital technologies (in particular, healthcare-related ones) without employing participatory approaches, and show that most of these issues can be overcome by, for instance, adapting the approach to the specific user group requirements, using metaphors and tangible objects to encourage and support envisioning of technology, and use non-technical language and providing explanations to avoid mismatched expectations.

4.3 Critical Corner

There is substantial evidence across HCI and ageing research pointing to methodological challenges in conducting research with older adults. Traditional research and design methods often fail to account for the wide range of sensory, physical, and cognitive characteristics associated with ageing (Newell et al., 2007). Age-sensitive methodological configurations are particularly important given that many older adults have not grown up with the same familiarity or fluency in using contemporary user interfaces and digital devices as younger generations (Rice et al., 2007). This claim, made several years ago, is still important, and it is likely to be so when most of today's adult people grow older too, as they will encounter digital technologies, they are not familiar with. As such, adaptations and adjustments are and will be needed across various components of the research and design process—including recruitment strategies, study materials, methodological approaches, participant engagement, and logistical planning. These configurations are essential for conducting research that is both respectful of older participants and capable of generating meaningful, high-quality insights for researchers.

On the one hand, the presence of methodological challenges when working with distinct participant groups—such as children, people with disabilities, and older adults—is to be expected. As the pool of research participants becomes increasingly diverse, it becomes increasingly unlikely that a single research method or activity will be universally effective. For instance, interviewing has traditionally followed a "one-size-fits-all" approach (Gubrium & Holstein, 2001, p. 179). While the push towards methodological standardization is understandable—aiming to guide researchers, facilitate comparisons, and promote high-quality studies (Silva et al., 2020)—this approach is increasingly recognized as problematic. As Gubrium and Holstein (2001, p. 179) argue, "the one-size-fits-all approach, especially as it applies to an ostensible category of respondents, is now less the operating rule than it is a significant procedural and analytic issue". In the light of this, methodological configurations are not only necessary but essential for conducting inclusive, context-sensitive research (see Fig. 4.3).

Fig. 4.3 Participant struggling with a questionnaire. "Why do you ask me to fill this questionnaire? Just ask me all the questions you want to know. It is difficult for people like us to understand all these questions and how to answer them. Am I being evaluated or what? It looks like the driving test I had to pass my get my driver's licence (smiling) a very long time ago"

On the other hand, it is perhaps more surprising—and even concerning—that research attention to methodological issues remains uneven within the field of older-adult HCI. Most of the studies reviewed in Sect. 4.2—excluding those focused on participatory design and co-design—are, or may be considered, outdated, having been published more than 10 years ago. This raises some interesting questions. To what extent are HCI researchers working with older adults today aware of the methodological challenges addressed in this body of knowledge? If a paper does not report on methodological configurations, does that imply the researchers encountered none? Is it due to contemporary digital technologies and contexts? When are methodological adaptations necessary, and why? Quantitatively, the literature discussed in this chapter represents less than 10% of the approximately 1000 works referenced throughout this book. While this is a rough estimate, and the synthesis presented here does not constitute a systematic literature review, it nonetheless suggests that a significant portion of HCI publications concerning older adults offer only a partial view of the research process and the methodological decisions involved. While some studies explicitly advocate for methodological adaptations and provide evidence of the changes, adjustments, or configurations implemented, others appear to overlook the need to tailor research methods and activities when working with older adults. This inconsistency suggests a gap in the field's collective awareness and reporting practices, despite the well-documented importance of age-sensitive methodological approaches.

Methodological configurations and reflections are a sign of the maturation of an academic field. Methodological issues are inherently complex, yet they play a fundamental role in shaping the quality and validity of research outcomes. Adaptations might not always be needed. Yet, the important point is that we need to know when, how, and why this happens, and to do so, further methodological transparency in publications would be useful, as it has been suggested in HCI. Little progress on methodological issues seems to have been made over the last two decades or so of older-adult HCI research. Whereas the literature reviewed in Sect. 4.2 covers a lot of ground, some research methods have not either been examined or at least received little research attention, such as remote usability testing (Hill et al., 2021a, 2021b).

References

Barbosa Neves, B., Franz, R., Judges, R., Beermann, C., & Baecker, R. (2019). Can digital technology enhance social connectedness among older adults? A feasibility study. *Journal of Applied Gerontology, 38*(1), 49–72. https://doi.org/10.1177/0733464817741369

Barrett, J., & Kirk, S. (2000). Running focus groups with elderly and disabled elderly participants. *Applied Ergonomics, 31*(6), 621–629. https://doi.org/10.1016/S0003-6870(00)00031-4

Bilecen, B., & Fokkema, T. (2022). Conducting Empirical Research With Older Migrants: Methodological and Ethical Issues. *The Gerontologist, 62*(6), 809–815. https://doi.org/10.1093/geront/gnac036

Binda, J., Wang, X., & Carroll, J. M. (2018). Recruiting Older Adults in the Wild: Reflections on Challenges and Lessons Learned from Research Experience. Proceedings of the 12th EAI International Conference on Pervasive Computing Technologies for Healthcare, 290–293. https://doi.org/10.1145/3240925.3240947

Boldi, A., Cho, S., Kou, Y., Rapp, A., & Birk, M. V. (2023). Methodological challenges, risks, and ethical implications in game research. In *Companion Proceedings of the annual symposium on computer-human interaction in play* (pp. 350–351). https://doi.org/10.1145/3573382.3616026

Chase, J.-A.D. (2013). Methodological Challenges in Physical Activity Research With Older Adults. *Western Journal of Nursing Research, 35*(1), 76–97. https://doi.org/10.1177/0193945911416829

Chen, Y., & Buckingham, L. (2025). Methodological Reflections for Qualitative Research with Older Migrants: Focusing on Chinese-heritage Grandparents in New Zealand. *International Journal of Qualitative Methods, 24*, 16094069251333248. https://doi.org/10.1177/16094069251333247

Cornet, V. P., Daley, C. N., Srinivas, P., & Holden, R. J. (2017). User-Centered Evaluations with Older Adults: Testing the Usability of a Mobile Health System for Heart Failure Self-Management. *Proceedings of the Human Factors and Ergonomics Society Annual Meeting, 61*(1), 6–10. https://doi.org/10.1177/1541931213601497

Da Silva Junior, D. P., Alves, D. D., Carneiro, N., Matos, E. D. S., Baranauskas, M. C. C., & Mendoza, Y. L. M. (2024). GranDIHC-BR 2025-2035-GC1: New theoretical and methodological approaches in HCI. In *Proceedings of the XXIII Brazilian symposium on human factors in computing systems* (pp. 1–31). https://doi.org/10.1145/3702038.3702054

Davidson, J. L., & Jensen, C. (2013). Participatory design with older adults: An analysis of creativity in the design of mobile healthcare applications. C&C.

Dickinson, A., Arnott, J., & Prior, S. (2007). Methods for human—Computer interaction research with older people. *Behaviour & Information Technology, 26*(4), 343–352. https://doi.org/10.1080/01449290601176948

Dickinson, A., Goodman, J., Syme, A., Eisma, R., Tiwari, L., Mival, O., & Newell, A. (2003). Domesticating technology. In-home requirements gathering with frail older people

Edlin-White, R., Cobb, S., Floyde, A., Lewthwaite, S., Wang, J., & Riedel, J. (2012). From Guinea Pigs to Design Partners: Working with Older People in ICT Design. In P. Langdon, J. Clarkson, P. Robinson, J. Lazar, & A. Heylighen (Eds.), Designing Inclusive Systems (pp. 155–164). Springer London. https://doi.org/10.1007/978-1-4471-2867-0_16

Eisma, R. (2003). Mutual Inspiration in the development of new technology for older people. In the Proceedings of INCLUDE 2003, London, UK

Eisma, R., Dickinson, A., Goodman, J., Syme, A., Tiwari, L., & Newell, A. F. (2004). Early user involvement in the development of information technology- related products for older people. *Universal Access in the Information Society, 3*(2), 131–140. https://doi.org/10.1007/s10209-004-0092-z

Ferlatte, O., Karmann, J., Gariépy, G., Frohlich, K. L., Moullec, G., Lemieux, V., & Hébert, R. (2022). Virtual Photovoice With Older Adults: Methodological Reflections during the COVID-19 Pandemic. *International Journal of Qualitative Methods, 21*, 16094069221095656. https://doi.org/10.1177/16094069221095656

Franz, R., Munteanu, C., Barbosa-Neves, B., & Baecker, R. (2015). Time to Retire Old Methodologies? Reflecting on Conducting Usability Evaluations with Older Adults. In Proceedings of the 17th International Conference on Human-Computer Interaction with Mobile Devices and Services Adjunct (MobileHCI '15). Association for Computing Machinery, New York, NY, USA, 912–915. https://doi.org/10.1145/2786567.2794303

Franz, R. L., Baecker, R., & Truong, K. N. (2018). "I knew that, I was just testing you": Understanding Older Adults' Impression Management Tactics During Usability Studies. *ACM Transactions on Accessible Computing, 11*(3), 1–23. https://doi.org/10.1145/3226115

Franz, R. L., Findlater, L., & Wobbrock, J. O. (2019). Just Ask Me: Comparing Older and Younger Individuals' Knowledge of Their Optimal Touchscreen Target Sizes. The 21st International ACM SIGACCESS Conference on Computers and Accessibility, 591–593. https://doi.org/10.1145/330 8561.3354615

Giannakos, M., Markopoulos, P., Hourcade, J. P., & Antle, A. N. (2022). 'Lots done, more to do': The current state of interaction design and children research and future directions. *International Journal of Child-Computer Interaction, 33*, Article 100469. https://doi.org/10.1016/j.ijcci.2022.100469

Gilbert, N. (2008). Researching Social Life. SAGE.

Goodman, J. (2003). Age-old Question(naire)s

Gubrium, J., & Holstein, J. (2001). *Handbook of Interview Research*. SAGE Publications.

Gubrium, J., & Sankar, A. (1994). Qualitative Methods in Aging Research. SAGE Publications

Hakobyan, L., Lumsden, J., & O'Sullivan, D. (2015). Participatory Design: How to Engage Older Adults in Participatory Design Activities. *International Journal of Mobile Human Computer Interaction, 7*(3), 78–92. https://doi.org/10.4018/ijmhci.2015070106

Hill, J. R., Brown, J. C., Campbell, N. L., & Holden, R. J. (2021a). Usability-In- Place—Remote Usability Testing Methods for Homebound Older Adults: Rapid Literature Review. *JMIR Formative Research, 5*(11), Article e26181. https://doi.org/10.2196/26181

Hill, J. R., Harrington, A. B., Adeoye, P., Campbell, N. L., & Holden, R. J. (2021b). Going Remote—Demonstration and Evaluation of Remote Technology Delivery and Usability Assessment With Older Adults: Survey Study. *JMIR mHealth and uHealth, 9*(3), Article e26702. https://doi.org/10.2196/26702

Holtzblatt, K., & Beyer, H. (2017). *Contextual design: Design for life*. Morgan Kaufmann.

Hone, K. S., & Lines, L. (2004). Eliciting user requirements with older adults: Lessons from the design of an interactive domestic alarm system. *Universal Access in the Information Society, 3*(2), 141–148. https://doi.org/10.1007/s10209-004-0094-x

Inglis, E. A., Szymkowiak, A., Gregor, P., Newell, A. F., Hine, N., Shah, P., Wilson, B. A., & Evans, J. (2003). Issues surrounding the user-centred development of a new interactive memory aid. *Universal Access in the Information Society, 2*(3), 226–234. https://doi.org/10.1007/s10209-003-0057-7

Jacko, J. (2012). The Human-Computer Interaction Handbook. Fundamentals, Evolving Technologies, and Emerging Applications (3rd ed.). CRC Press

Lazar, J., Heidi, J., & Hochheiser, H. (2017). *Research Methods in Human- Computer Interaction.* Morgan Kaufmann.

Lee, H. R., Cheon, E., Lim, C., & Fischer, K. (2022). Configuring Humans: What Roles Humans Play in HRI Research. 2022 17th ACM/IEEE International Conference on Human-Robot Interaction (HRI), 478–492. https://doi.org/10.1109/HRI53351.2022.9889496

Lehnert, F. K., Niess, J., Lallemand, C., Markopoulos, P., Fischbach, A., & Koenig, V. (2022). Child-Computer Interaction: From a systematic review towards an integrated understanding of interaction design methods for children. *International Journal of Child-Computer Interaction, 32*, Article 100398. https://doi.org/10.1016/j.ijcci.2021.100398

Lindsay, S., Jackson, D., Schofield, G., & Olivier, P. (2012). Engaging older people using participatory design. In *Proceedings of the SIGCHI conference on human factors in computing systems* (pp. 1199–1208). https://doi.org/10.1145/2207676.2208570

Lofland, J., Snow, D., Anderson, L., & Lofland, L. (2006). *Analizing Social Settings.* Wadsworth.

Lörinc, M., Kilkey, M., Ryan, L., & Tawodzera, O. (2022). "You Still Want to Go Lots of Places": Exploring Walking Interviews in Research With Older Migrants. *The Gerontologist, 62*(6), 832–841. https://doi.org/10.1093/geront/gnab152

Marzi, K., Klapperich, H., & Huldtgren, A. (2023). Insights on Older Adults' Willingness, Motives and Experiences regarding Participation in HCI research. *Mensch und Computer, 2023*, 432–436. https://doi.org/10.1145/3603555.3608540

Massimi, M., Baecker, R. M., & Wu, M. (2007). Using participatory activities with seniors to critique, build, and evaluate mobile phones. Proceedings of the 9th International ACM SIGACCESS Conference on Computers and Accessibility, 155– 162. https://doi.org/10.1145/1296843.1296871

Mathiesen, S. L., Grenier, A., Wittich, W., Sukhai, M., & Herrmann, B. (2025). Narrative Engagement in Story Listening: The Challenge of Age and Vision Loss. Journal of Aging Studies. https://doi.org/10.31235/osf.io/rgm5w

Mehta, K. K. (2011). The challenges of conducting focus-group research among Asian older adults. *Ageing and Society, 31*(3), 408–421. https://doi.org/10.1017/S0144686X10000930

Meissner, J. L. (2025). Configuring Participatory Research as Give and Take Relationships: Methodological Reflections on Co-Designing Booklets with a Men Shed. Proceedings of the 2025 CHI Conference on Human Factors in Computing Systems, 1–17. https://doi.org/10.1145/3706598.3713106

Minocha et al. (2013).Conducting Empirical Research with older Adults. CHI

Nathan, P. (2014). *The Oxford Handbook of Qualitative Research.* Oxford University Press.

Newell, A., Arnott, J., Carmichael, A., & Morgan, M. (2007). Methodologies for Involving Older Adults in the Design Process. In C. Stephanidis (Ed.), Universal Acess in Human Computer Interaction. Coping with Diversity (Vol. 4554, pp. 982– 989). Springer Berlin Heidelberg. https://doi.org/10.1007/978-3-540-73279-2_110

Olson, J., & Kellog, W. (2014). *Ways of Knowing in HCI.* Springer.

Oppenlaender, J., Malacria, S., Fang, X., Van Berkel, N., Chevalier, F., Yatani, K., et al. (2025). Meta-HCI: First workshop on meta-research in HCI. In *Proceedings of the extended abstracts of the CHI conference on human factors in computing systems* (pp. 1–8). https://doi.org/10.1145/3706599.3706723

Park, D., & Schwarz, N. (2000). *Cognitive aging.* Taylor & Francis.

Petrie, H. (2025). The ethics of conducting research with human participants in HCI. In Vanderdonckt, J., Palanque, P., & Winckler, M. (Eds.), *Handbook of human computer interaction.* Springer, Cham. https://doi.org/10.1007/978-3-319-27648-9_86-1

Platzer, F., Steverink, N., Haan, M., De Greef, M., & Goedendorp, M. (2021). The bigger picture: Research strategy for a photo-elicitation study investigating positive health perceptions of older adults with low socioeconomic status. *International Journal of Qualitative Methods, 20,* 16094069211040950. https://doi.org/10.1177/16094069211040950

Pradhan, A., Jelen, B., Thangaraj, R., Siek, K. A., Jette, S., & Lazar, A. (2025). Understanding older adults' (Dis)engagement with design materials. In *Proceedings of the 2025 CHI conference on human factors in computing systems* (pp. 1–17). https://doi.org/10.1145/3706598.3713846

Qi, X., & Yu, J. (2025). *Participatory design in human-computer interaction: Cases, characteristics, and lessons.* CHI.

Quinn, K. (2010). Methodological considerations in surveys of older adults. *Technology Matters, 8*(2).

Regalado, F., Santos, C., & Veloso, A. I. (2025). Co-creating with older adults is a challenging task! Lessons learned from real-life experiences and challenges at an ageing lab. In *Proceedings of the 2025 ACM international conference on interactive media experiences* (pp. 139–147). https://doi.org/10.1145/3706370.3727865

Rice, M., Newell, A., & Morgan, M. (2007). Forum theatre as a requirements gathering methodology in the design of a home telecommunication system for older adults. *Behaviour & Information Technology, 26*(4), 323–331. https://doi.org/10.1080/01449290601177045

Robertson, J., & Kaptein, M. (2016). *Modern statistical methods for HCI.* Springer.

Rogers, Y., Sharp, H., Benyon, D., Holland, S., & Carey, T. (1994). *Human-computer interaction.* Addison-Wesley.

Rubin, J. (1994). Handbook of usability testing. In *How to plan, design, and conduct esective tests.* John Wiley & Sons.

Salomé, S., Du Bousquet, L., & Monfort, E. (2025). Co-designing conversational agents with older people: A scoping review of methods, challenges, and a way forward. *Computers in Human Behavior Reports, 17,* Article 100606. https://doi.org/10.1016/j.chbr.2025.100606

Sayago, S (2023). Cultures in human-computer interaction. In *Synthesis lectures on human-centered informatics.* Springer Cham.

Sayago, S., & Blat, J. (2011). An ethnographical study of the accessibility barriers in the everyday interactions of older people with the web. *Universal Access in the Information Society, 10*(4), 359–371.

Sharp, H., Preece, J., & Rogers, Y. (2023). *Interaction design: Beyond human- computer interaction (6th ed.).* Wiley.

Shoenfelt, A., Isaacowitz, D. M., & Ebner, N. C. (2025). Eye tracking in aging: Challenges, best practices, and novel frontiers. *ACM Transactions on Applied Perception, 3733834.* https://doi.org/10.1145/3733834

Silva, A., Isabel Martins, A., Caravau, H., Margarida Almeida, A., Silva, T., Ribeiro, Ó., Santinha, G., et al. (2020). Experts evaluation of usability for digital solutions directed at older adults: A scoping review of reviews. In *Proceedings of the 9th international conference on software development and technologies for enhancing accessibility and fighting info-exclusion* (pp. 174–181). https://doi.org/10.1145/3439231.3439238

Simonsen, J., & Robertson, T. (2013). *Routledge international handbook of participatory design.* Routledge.

Trujillo Tanner, C., Caserta, M. S., Kleinschmidt, J. J., Clayton, M. S., Bernstein, P. S., & Guo, J. W. (2018). Conducting research with older adults with vision impairment: Lessons learned and recommended best practices. *Gerontology and Geriatric Medicine, 4*, 2333721418812624. https://doi.org/10.1177/2333721418812624

Vines, J., Blythe, M., Dunphy, P., Vlachokyriakos, V., Teece, I., Monk, A., et al. (2012). Cheque mates: Participatory design of digital payments with eighty somethings. In *Proceedings of the SIGCHI conference on human factors in computing systems* (pp. 1189–1198). https://doi.org/10.1145/2207676.2208569

Waycott, J., Davis, H., Thieme, A., Branham, S., Vines, J., & Munteanu, C. (2015). Ethical encounters in HCI: research in sensitive settings. In *Proceedings of the 33rd annual ACM conference extended abstracts on human factors in computing systems* (pp. 2369–2372). https://doi.org/10.1145/2702613.2702655

Weil, J. (2017). *Research design in ageing and social gerontology, quantitative, qualitative, and mixed methods.* Routledge.

Wiberg, M. (2026). *Fast and slow.* CRC Press.

Looking Ahead 5

Abstract

This chapter wraps up the book by looking ahead. If a whole society shifts to longer lives, what happens to definitions of later life and ageing? How will this greater engagement in older age affect digital technology conceptualisation, design, and use? Where will a repositioning of older adults within a broader ecosystem of more-than-human ageing shape older-adult HCI in the future? How can older-adult HCI research contribute to post-growth development goals (e.g. reduced inequalities, gender equality, sustainable cities, climate action)? This chapter discusses these and other related research questions as an invitation to reflect on possible future research directions.

Keywords

Culture shift · Synthetic research · More-than-human ageing · Knowledge production

5.1 Purpose and Context

As stated in the introductory chapter, an ageing population provides HCI scholars with the rare opportunity to study how HCI responds to a fundamentally new and unprecedented phenomenon. This chapter is an invitation to reflect further on possible future research directions that the field of older-adult HCI could explore. These future perspectives are inspired by the previous chapters and recent or ongoing developments in HCI and digital technologies, which are discussed throughout this chapter.

S. Sayago, *Older Adults and Digital Technologies*, Synthesis Lectures on Human-Centered Informatics, https://doi.org/10.1007/978-3-032-25556-3_5

5.2 Culture Shift

As discussed by Carstensen (2015), the culture that guides most people through life today is a culture that evolved around shorter lives. The demographic changes underway are fundamentally altering virtually all aspects of life as we know it, from workforces to families and education. In her view, the urgent challenge now is to create cultures that support people through ten or more decades of life and to think strategically about how to best use these extra years.

HCI can also be seen as constrained by cultures (significant plural) (Sayago, 2023) that evolved around shorter lives. There is widespread agreement in the HCI literature that most digital technologies have been designed by overlooking older adults. Usability concerns have for too long overshadowed questions about the usefulness and acceptability of digital technologies for older adults (Knowles et al., 2021). Methodological challenges involved when working with older research participants indicate that what works for "ordinary" HCI users might not work for "extraordinary" user groups like older adults. Older people and ageing, despite gaining traction in HCI, do not play a prominent role in accessibility research, which focuses heavily on the needs of blind and low vision users (Mack et al., 2021).

If we are going to live longer lives, as demographics suggest, and we agree on ageing being a process of change, as discussed in Chap. 1, we might need to think about designing digital technologies that "grow" older along with us. Further research could also be conducted with centenarians, who are gaining visibility in demographics, and other members of the "oldest-old", to gain a deeper and broader understanding of ageing within HCI and the role(s) of digital technologies in their lives. Despite some exemptions (Barbosa-Neves et al., 2015), these older adults do not feature prominently in older-adult HCI research. As discussed in Chap. 4, design and research methods and activities might also need to be re-configured to accommodate and include older people in our work.

If a whole society shifts to live longer, what happens to definitions of later life and ageing? How will this greater engagement in older age affect digital technology conceptualisation, design, and use?

These and other proposed directions come down to keep configuring (i.e. cultural shift) HCI to respond to a demographic reality and support people—by means of digital technologies—through ten and more decades of life and to think strategically about how to best use HCI knowledge to live longer and richer hybrid lives. To know more about ageing and older adults is to know more about ourselves.

5.3 More-Than-Human Ageing

As discussed by Coskun et al. (2022), we are witnessing the expansion of the design space from a human-centred perspective to more-than-human design orientations. Two main developments are driving the rise of more-than-human design. The first stems from the Anthropocene (Lewis & Maslin, 2015) and the recognition that reality extends beyond human experience (Norman, 2023). This perspective acknowledges the roles of non-human actors, such as animals, environmental systems (including those affected by the current ecological crisis), and technologies in shaping our world (Giaccardi et al., 2025). The second development is the growing recognition that contemporary digital technologies possess forms of agency and capacity for action that far exceed those of earlier technological systems (Cila et al., 2017; Giaccardi et al., 2025). Their influence is so substantial that it may be misleading to conceptualise our relations with them merely in terms of "use" (Baumer & Brubaker, 2017) or "interaction" (Frauenberger, 2020; Verbeek, 2015). As digital technologies have gained greater computational power and network connectivity, many of them (e.g. IoT, voice assistants) have become actants with performative roles in our lives (Cila et al., 2017). Technology is not just material but participant (Giaccardi & Redström, 2020). This situation expands the scope of inquiry from a single archetypical user with an single product to multiple products, services, and stakeholders with various roles and relations to each other, which in turn, questions the adequacy of human-centred design (Coskun et al., 2022; Giaccardi & Redström, 2020; Giaccardi et al., 2025). As such, more-than-human perspectives are gaining ground in HCI research (Eriksson et al., 2024).

Within older-adult HCI, understanding what older adults want or need and designing technologies to ensure the best possible outcome and user experience is the core of what is often referred to as good (human-centred) design with this group. Chapter 3 shows that designing digital technologies for older adults is a challenging task. Embracing more-than-human design orientations in older-adult HCI might not make it less challenging. Yet, it can help us move past the potential limitations of human-centred design at the outset of the twenty-first century and address other elements that are entering design practice with older people too. For example, the reality of older adults considered in digital technology design could consider more than just their human experience to include other non-human actants. Recent work in critical gerontology is beginning to question whether a sole humanistic focus can provide a holistic perspective on technologies for ageing (Pradhan et al., 2025). As stated by Lupton (2024), none of us ages alone. To what extent are most older adults aware of the influence of the expanding universe of algorithms and digital forms of intelligence in their lives? How do these non-human actants relate or should relate to them?

Most of the discussion on technology design for and with older people (see Chap. 3) focuses on short-term needs and goals. For instance, digital technologies are designed to enable older adults to do something (e.g. send an e-mail or share a photo online) or achieve important objectives for them (and us), such as avoiding social isolation and/or exclusion. It could also be worth thinking about the social and long-term environmental consequences (Sharma et al., 2024) of designing technologies that foster positive (or "negative") ageing, as designing technology is designing human beings and human existence (Norman, 2023; Verbeek, 2015). To what extent are most of the technologies designed for older adults feeding or reducing the fourth age or stereotypes? What does the next generation of older people (i.e. today's adult people) think about these technologies? How do these developments relate to post-growth (Sharma et al., 2024) development goals (e.g. reduced inequalities, gender equality, sustainable cities, climate action)?

Post-userism might provide us with a way to move towards a more comprehensive account of the relationships that occur between older people and computer-enabled technologies that sit outside the bounds of what we would likely refer to as "use", such as adult children making online appointments to the doctor or conducting online banking tasks on behalf of their older parents. What other configurations or mediations between digital technologies and older people are taking place or should be designed for? What are their implications in terms of technology use, design, and uptake?

To sum up, where will a repositioning of older adults within a broader ecosystem (of human and non-human actants) take us and how will it shape older-adult HCI in the future?

5.4   Tomorrow's Older Adults

5.4.1   Flesh-and-Blood Individuals

As obvious as it may be, a great deal of research is conducted with contemporary older adults. Prompted by a seminal work of Hanson (2009), an important question that remains to be addressed is how what we know nowadays about older adults and digital technologies in HCI will change when today's adult people grow older (i.e. the next generation of older people).

Addressing this question might aid in distinguishing between time-persistent issues, which are unlikely to change in significant ways in the future, and others more related to technology, which are expected to change due to the constant evolution of technologies. For instance, Sayago et al. (2011) suggest that cognition is always more important than vision in older people's interaction with digital technologies, regardless of their experience with computers, as vision-related issues are solved with technologies (reading glasses) and experience with technologies (zoom), whereas cognitive-related ones are time-persistent (e.g. difficulties in learning new things, remembering how to do a task after a period of inactivity) and do not seem to have been solved by technologies yet.

A growing diversity is expected in the next wave of older people, due to the third age, consumerism, migration, and increasingly varied lifestyles. Older adults today are different in many dimensions as compared to prior generations. Generally, the level of education of the current generation of older adults is higher than that of previous generations, and this trend will increase in the future with more people becoming college-educated beyond high school (Czaja & Weingast, 2020). Although most older adults may be positively disposed towards the idea of using new technology in the future, new is always new (Rogers & Mitzner, 2017), and experience with today's technology may not transfer easily (or at all) to use of tomorrow's technology (Sayago et al., 2013). With new generations of older people, comes new or different digital divides.

For example, unless technological development slows the technological divide, today's technologies will be replaced by new ones (Cabrera & Malanowski, 2009), which might bring about a new or different digital divide. Access to technology, which has been (and currently is) a relevant barrier for most of today's older people, is likely to be much less important in the future than algorithmic, robot, and AI literacy (Rasi-Heikkinen et al., 2024; Zhou et al., 2025).

How will the next wave(s) of older adults, and new technological developments, change the nature and understanding of older-adult computer interaction research and design?

5.4.2 Synthetic Individuals

Looking ahead, it might be worth exploring Large Language Models (LLMs) and generative artificial intelligence (GenAI) with older adults. GenAI is a hot topic in HCI, e.g. (Capel & Brereton, 2023; Liu et al., 2025; Shneiderman, 2022; Zhou et al., 2025). In keeping with the type of knowledge production in older-adult HCI, which is largely determined by technological innovations (see Section 5.5), there are reasons to believe that a future wave of studies about GenAI and older people will become available soon. In fact, some of these "future" works have already been published at the time of writing this book. For example, Sheahan et al. (2024) address the question of how we can utilise GenAI to co-design fictions that promote more desirable everyday futures for older people. Liu et al. (2025) explore LLM-powered voice gents that support interruptions in conversations with older people. Enam et al. (2025) examine older adults' perceptions towards LLM-based chatbots, and Zhai et al. (2025) address the interesting topic of deepfakes with older adults.

This presents us with an opportunity to explore issues that are receiving growing research attention in HCI with older people so that they are not overlooked—as it tends to happen, especially at the beginning of many technological revolutions. For example, to what extent do guidelines for human–AI interaction (Amershi et al., 2019) relate to older adults? The current hype surrounding GenAI also enables us to delve into issues that matter in the interactions (or mediations) between older people and digital technologies, from trust to control, in ways that were difficult to do until now, by augmenting human and designer researchers' activities and capabilities with GenAI systems. How can humans teaming with AI (Xu & Gao, 2024) aid in designing more trustworthy technologies for older people? Assuming that meaningful synthetic older people can be created and simulated with existing datasets,[1] how can these interactive personas contribute to designing better technologies for older people? How can GenAI help us to build digital technologies that grow older with us? Ongoing discussions on replacing human participants in technology development and scientific research with LLM are taking place. As discussed by Agnew et al. (2024), these proposals are motivated by reducing costs, protecting participants from potential harms, and augmenting diversity. However, these proposals undermine empowerment and understanding. Future research could explore this "illusion of artificial inclusion", as discussed by Agnew et al. (2024), with older adults.

These and other questions might be a starting point for a discussion in the community about how we want to advance this field in an era of GenAI to create more usable, accessible, useful, inclusive and less discriminatory interactive experiences for and with older adults (and the rest of us).

5.5   Knowledge Production: Breadth and Depth

Older-adult HCI research tends to mirror the latest technological developments. When a new technology is introduced, it tends to generate a flow of papers exploring its relationship with older people. Most of the previous technologies fade away or are rendered invisible. Table 5.1 illustrates this fact.[2]

This interest in addressing contemporary technologies might be expected, or regarded as natural, in a field like HCI, which strives to keep up with the rapid pace of technological advancement (Wiberg, 2026). However, these repeated shifts in emphasis have implications that deserve consideration.

To begin with, this type of knowledge production, which is strongly driven by technological developments, might potentially lead some of us to believe that we already know how to design certain types of technologies for older adults. In other words, if research contributions do not address technologies such as desktop computers or ATMs, it means we already know enough about the relationship between older adults and them. Otherwise, research would be focused on these technologies. Do we truly know how to

---

[1] Masoli et al., (2023) and Park et al., (2021) show that older adults are "under-represented" in several datasets and Rosales and Fernández-Ardévol (2019) discuss structural ageism in big data approaches.

[2] The source of Table 5.1 is the Bibliography available at the end of the book. A decade has been used as a temporal unit to provide a snapshot of the evolution of digital technologies that have received research attention in the field over time.

Table 5.1 Knowledge production: an overview

Year	A snapshot of digital technologies that receive research attention
2021—year of writing this book (2025)	Mobile applications, voice assistants, wearables, virtual and augmented reality, online banking, makers, social media and online social networks, smart devices, social robots, chatbots, voice assistants, virtual environments, automated vehicles
2011–2020	Mobile phones and devices, online social networks and communities, tablets, care and socially assistive robots, virtual companions, digital games, smartphones, voice assistants, social media, virtual agents, making or DIY technologies
2000–2010	Computers, the web, tablets, instant messaging, gestural interfaces, surface computing, mobile phones, social media, instant messaging, touch-based interfaces
1990–2000	Computers, word processing, automatic teller machines, websites

design websites, desktop computers, word processing tools and mobile apps for older adults? Are videoconferencing systems designed by considering them? Throughout the COVID-19 pandemic, it became evident that most online services and computer-mediated communication technologies still present usability and accessibility issues for older adults (Tanprasert et al., 2024). This is not to say that they are unable to use these technologies. However, this example shows that 'old' technologies are still far from being "older people" friendly, and little research is currently focusing on them. The current climate of *publish or perish*, and the rise and fall of digital technologies (e.g. it is rare to find research studies about Facebook in the HCI literature nowadays), might account for this type of research knowledge production.[3]

Another implication is that the creation of a body of knowledge on which we can draw to design different types of technologies is not ready yet. Whereas exploring different technologies allows us to get a better understanding of issues that matter to older adults, the field seems to start from scratch with every new technology.[4] For example, there is widespread agreement in the literature that most older adults need to be in control of digital technologies (Sánchez Chamorro et al., 2024), from home robotic devices (Deutsch et al., 2019) and health monitoring technologies (Berridge et al., 2022) to assistive technologies (Ogonowski et al., 2016; Yusif et al., 2016) and social media (Xie et al., 2012). We know that taking control matters. Research with different technologies

---

[3] This type of knowledge production is not unique to older-adult HCI. It has also been found in children-computer interaction (Torgersson et al., 2019).

[4] A similar situation is discussed in mainstream HCI research (Liu et al., 2014).

repeats this message. Yet, how do we use the knowledge accumulated over these years to design for the control that most older people need or look for in new or emerging digital technologies? I do not have a clear answer.

Privacy and trust are another example. Much evidence shows that privacy and trust are critical issues for most older adults to accept, adopt, and use digital technologies, e.g. (Aziz et al., 2025; Bonilla & Martin-Hammond, 2020; Chattaraman et al., 2012; Ghaiumy Anaraky et al., 2021; Harrington & Egede, 2023; Hope et al., 2014; Hornung et al., 2017; Joshi et al., 2025; Latulipe et al, 2022; Lee & Coughlin, 2015; Montuwy et al., 2018). Privacy has been identified as one of the most important barriers to adoption, even more than usability (Zhou et al., 2025). Yet, and despite some exceptions, which focus on (dis)trust and privacy (Ghaiumy Anaraky et al., 2021; Harrington & Egede, 2023; Knowles & Hanson, 2018a, 2018b; Zhang et al., 2024; Zhou et al., 2025), these issues tend to be mentioned in passing, i.e. without delving into them. For example, "current research has not addressed the privacy-specific concerns of older adults" (Zhang et al., 2024). However, a quick online search for trust or privacy in Google Scholar, SCOPUS and Web of Science, for example, shows that the academic literature (in general, not specific to HCI) about these two topics is massive.

How can we extend what is known about trust and privacy in, for instance, voice assistants and online banking, to virtual influencers (Kugler, 2023) or AI-powered chatbots (Aziz et al., 2025), and vice versa? Which things change and why? It is hard to find an answer in the literature. The discussion on privacy (and trust too) is a complex one that goes beyond technological aspects (Zhou et al., 2025). Also, knowledge in HCI, and in older-adult HCI, tends to be highly contextual (Liu et al., 2014), due to the rapid pace of technology designed for humans. An exemption is knowledge about ageing, which does not change as rapidly as technology. Yet, addressing these and related questions is important to strike a better balance between research driven by technological developments, which is undoubtedly valuable in today's fast-paced digital environment, and research that addresses issues that play an important role in technology use by older adults, and that demands further reflection and comprehensive understanding to make a long-lasting impact. As discussed by Wiberg (2026), not all research problems share the same temporal grounds. It depends on the type of problem. It might not be easy, but I consider that there is a need to find a balance so that we can delve further into those issues that matter to most older adults—without overlooking emerging technological developments—to both keep taking the field forward and to develop further cooperative research themes across technologies.

5.6   The Impact of Digital Technologies

With each new digital technology comes great enthusiasm about its potential positive effects on the everyday lives of older adults. From socially assistive robots, and virtual and augmented reality, to voice assistants, smartphones and tablets, assistive technologies and smart homes, there is a widespread belief that digital technologies may (or should) play a key and positive role in solving older people's problems and improving their quality of life and subjective well-being.

However, despite decades of research involving different profiles of older adults and digital technologies, e.g. (Blaschke et al., 2009; Bobillier Chaumon et al., 2014; Camargo et al., 2025; Castro Martínez et al., 2025; Czaja et al., 2024; D'Haeseleer et al., 2020; Dickinson & Gregor, 2006; Jin et al., 2024; Kachouie et al., 2014; Khosravi et al., 2016; Liu et al., 2016; Parra et al., 2014; Thach et al., 2020; Wilson et al., 2022), there is inconclusive and mixed evidence of the impact of digital technologies on enhancing the quality of life, subjective well-being, and independence of many older adults.

This is not to say that digital technologies do not have an impact on the lives of older people. There is a need to separate the evidence base for claims from utopian perceptions or overly optimism about the ultimate value of digital technologies. The impact of digital technologies is not universally positive or negative. It depends on individual circumstances and broader social contexts. Also, quality of life and well-being are complex topics. The factors that play a key role in technology acceptance and use by older people are so many and diverse (Lee & Coughlin, 2015) that it is unreasonable to think that digital technologies will always have a positive or negative impact on their subjective well-being and quality of life. Outcomes vary depending on the individual, the specific technology, the training provided, socio-cultural context, education, social support, perceived usefulness and ease of use, and even patterns of use (Kraut & Burke, 2015; Lifshitz et al., 2018). At the same time there is a need for more robust evaluations of these technologies, involving larger samples, and more longitudinal, rigorous study designs (Kraut & Burke, 2015). Nowadays, coherent and unequivocal empirical evidence of the potential of digital technologies to improve older people's lives seems to be "infinitely postponed" (Dourish & Bell, 2011).

Looking ahead, Peine and Neven (2019) suggest that instead of focusing on impact as qualities that can be defined a priori.[5] Studies in ageing and technology could explore how this impact is created in interaction with technology, as ageing and technology are co-constituted, e.g. telecare technologies not only support living longer at home but change the very experience of the home itself.

Another possibility, given that much research is driven by technological developments and their assumed potential, could be to focus more on what older adults really want and/or need (Ghorayeb et al., 2021). Instead of asking whether older adults are able and

---

[5] Such as UTAUT2 (Unified Theory of Acceptance and Use of Technology), which has proven to be a relevant theoretical base to explain ICT use among older people (Macedo, 2017).

willing to accept, adopt and use emerging digital technologies, which are often imposed on them, and then measure their impact on their quality of life, could or should we ask ourselves more often whether mainstream and emerging digital technologies are ready to be accepted—adopted[6] and used—by older adults, and if not, what changes should be made in the technologies?

Previous research shows that older adults are highly selective users of technologies (Knowles & Hanson, 2018a, 2018b; Sayago et al., 2011). Therefore, technologies that take older adults' concerns, interests, needs, and use patterns into account (Lifshitz et al., 2018) might offer a richer understanding of how digital technologies affect—or fail to affect—their lives and, by extension, ours.

References

Agnew, W., Bergman, A. S., Chien, J., Díaz, M., El-Sayed, S., Pittman, J., Mohamed, S., & McKee, K. R. (2024). The illusion of artificial inclusion. In *Proceedings of the CHI conference on human factors in computing systems* (pp. 1–12). https://doi.org/10.1145/3613904.3642703

Amershi, S., Weld, D., Vorvoreanu, M., Fourney, A., Nushi, B., Collisson, P., Suh, J., Iqbal, S., Bennett, P. N., Inkpen, K., Teevan, J., Kikin-Gil, R., & Horvitz, E. (2019). Guidelines for human-AI interaction. In *Proceedings of the 2019 CHI conference on human factors in computing systems* (pp. 1–13). https://doi.org/10.1145/3290605.3300233

Aziz, F., Law, E. L.-C., Li, L., & Chen, S. (2025). Inclusive AI-driven music Chatbots for older adults. In L. Zaina, J. C. Campos, D. Spano, K. Luyten, P. Palanque, G. Van Der Veer, A. Ebert, S. R. Humayoun, & V. Memmesheimer (Eds.), *Engineering interactive computer systems. EICS 2024 international workshops* (Vol. 15518, pp. 74–93). Springer. https://doi.org/10.1007/978-3-031-91760-8_6

Barbosa-Neves, B., Franz, R., Munteanu-C., Baecker, R. & Ngo, M. (2015). "My hand doesn't listen to me!": Adoption and evaluation of a communication technology for the 'oldest old'. In *Proceedings of the 33rd annual ACM conference on human factors in computing systems (CHI'15)* (pp. 1593–1602). Association for Computing Machinery. https://doi.org/10.1145/2702123.2702430

Baumer, E. P. S., & Brubaker, J. R. (2017). Post-userism. In *Proceedings of the 2017 CHI conference on human factors in computing systems* (pp. 6291–6303). https://doi.org/10.1145/3025453.3025740

Berridge, C., Zhou, Y., Lazar, A., Porwal, A., Mattek, N., Gothard, S., & Kaye, J. (2022). Control matters in elder care technology: Evidence and direction for designing it in. In *Designing interactive systems conference* (pp. 1831–1848). https://doi.org/10.1145/3532106.3533471

Blaschke, C. M., Freddolino, P. P., & Mullen, E. E. (2009). Ageing and technology: A review of the research literature. *British Journal of Social Work, 39*(4), 641–656. https://doi.org/10.1093/bjsw/bcp025

Bobillier-Chaumon, M.-E., Michel, C., Tarpin-Bernard, F., & Croisile, B. (2014). Can ICT improve the quality of life of elderly adults living in residential home care units? From actual impacts

---

[6] To study computer use after initial adoption (Macedo, 2017) argues that popular constructs in technology acceptance, such as perceived ease-of-use, offer limited value.

to hidden artefacts. *Behaviour and Information Technology, 33*(6), 574–590. https://doi.org/10.1080/0144929X.2013.832382

Bonilla, K., & Martin-Hammond, A. (2020). *Older adults' perceptions of intelligent voice assistant privacy, transparency, and online privacy guidelines.*

Cabrera, M., & Malanowski, N. (Eds.). (2009). *Information and communication technologies for active ageing: Opportunities and challenges for the European Union.* IOS Press.

Camargo, J., Silva, T., & Ferraz De Abreu, J. (2025). Connection between the real world and the digital world: Voice assistants as promoters of socialization for older adults. In *Proceedings of the 11th international conference on information and communication technologies for ageing well and e-health* (pp. 406–413). https://doi.org/10.5220/0013501100003938

Capel, T., & Brereton, M. (2023). What is human-centered about human-centered AI? A map of the research landscape. In *Proceedings of the 2023 CHI conference on human factors in computing systems* (pp. 1–23). https://doi.org/10.1145/3544548.3580959

Carstensen, L. (2015) *The future of ageing.* Futureagenda.org

Castro-Martínez, E., Hernández-Encuentra, E., & Pousada-Fernández, M. (2025). Voice assistants' influence on loneliness in older adults: A systematic review. *Disability and Rehabilitation: Assistive Technology, 20*(3), 521–535. https://doi.org/10.1080/17483107.2024.2397030

Chattaraman, V., Kwon, W.-S., & Gilbert, J. E. (2012). Virtual agents in retail web sites: Benefits of simulated social interaction for older users. *Computers in Human Behavior, 28*(6), 2055–2066. https://doi.org/10.1016/j.chb.2012.06.009

Cila, N., Smit, I., Giaccardi, E., & Kröse, B. (2017). Products as agents: Metaphors for designing the products of the IoT age. In *Proceedings of the 2017 CHI conference on human factors in computing systems* (pp. 448–459). https://doi.org/10.1145/3025453.3025797

Coskun, A., Cila, N., Nicenboim, I., Frauenberger, C., Wakkary, R., Hassenzahl, M., Mancini, C., Giaccardi, E., & Forlano, L. (2022). More-than-human concepts, methodologies, and practices in HCI. In *CHI conference on human factors in computing systems extended abstracts* (pp. 1–5). https://doi.org/10.1145/3491101.3516503

Czaja, S. J., Charness, N., Rogers, W. A., Sharit, J., Moxley, J. H., & Boot, W. R. (2024). The benefits of technology for engaging aging adults: Findings from the PRISM 2.0 trial. *Innovation in Aging, 8*(6), igae042. https://doi.org/10.1093/geroni/igae042

Czaja, S. J., & Weingast, S. Z. (2020). The changing face of aging: Characteristics of older adult user groups. *Gerontechnology, 19*(2), 115–124. https://doi.org/10.4017/gt.2020.19.2.004.00

D'Haeseleer, I., Oeyen, D., Vanrumste, B., Schreurs, D., & Abeele, V. (2020). The unacceptance of a self-management health system by healthy older adults. In *Proceedings of the 6th international conference on information and communication technologies for ageing well and e-health* (pp. 269–280). https://doi.org/10.5220/0009806402690280

Deutsch, I., Erel, H., Paz, M., Hoffman, G., & Zuckerman, O. (2019). Home robotic devices for older adults: Opportunities and concerns. *Computers in Human Behavior, 98*, 122–133. https://doi.org/10.1016/j.chb.2019.04.002

Dickinson, A., & Gregor, P. (2006). Computer use has no demonstrated impact on the well-being of older adults. *International Journal of Human-Computer Studies, 64*(8), 744–753. https://doi.org/10.1016/j.ijhcs.2006.03.001

Dourish, P., & Bell, G. (2011). *Divining a digital future: Mess and mythology in ubiquitous computing.* The MIT Press

Enam, M. A., Murmu, C., & Dixon, E. (2025). "Artificial intelligence—carrying us into the future": A study of older adults' perceptions of LLM-based Chatbots. *International Journal of Human-Computer Interaction, 64*, 1–24. https://doi.org/10.1080/10447318.2025.2476710

Eriksson, E., Yoo, D., Bekker, T., & Nilsson, E. M. (2024). More-than-human perspectives in human-computer interaction research: A scoping review. *Nordic Conference on Human-Computer Interaction, 24*, 1–18. https://doi.org/10.1145/3679318.3685408

Frauenberger, C. (2020). Entanglement HCI the next wave? *ACM Transactions on Computer-Human Interaction, 27*(1), 1–27. https://doi.org/10.1145/3364998

Ghaiumy-Anaraky, R., Byrne, K. A., Wisniewski, P. J., Page, X., & Knijnenburg, B. (2021). To disclose or not to disclose: Examining the privacy decision-making processes of older versus younger adults. In *Proceedings of the 2021 CHI conference on human factors in computing systems* (pp. 1–14). https://doi.org/10.1145/3411764.3445204

Ghorayeb, A., Comber, R., & Gooberman-Hill, R. (2021). Older adults' perspectives of smart home technology: Are we developing the technology that older people want? *International Journal of Human–computer Studies, 147*, 102571. https://doi.org/10.1016/j.ijhcs.2020.102571

Giaccardi, E., & Redström, J. (2020). Technology and more-than-human design. *Design Issues, 36*(4), 33–44.

Giaccardi, E., Redström, J., & Nicenboim, I. (2025). The making(s) of more-than-human design: Introduction to the special issue on more-than-human design and HCI. *Human-Computer Interaction, 40*(1–4), 1–16. https://doi.org/10.1080/07370024.2024.2353357

Hanson, V. L. (2009). Age and web access: The next generation. In *Proceedings of the 2009 international cross-disciplinary conference on web accessibililty (W4A)* (pp. 7–15). https://doi.org/10.1145/1535654.1535658

Harrington, C. N., & Egede, L. (2023). Trust, comfort and relatability: Understanding black older adults' perceptions of Chatbot design for health information seeking. In *Proceedings of the 2023 CHI conference on human factors in computing systems* (pp. 1–18). https://doi.org/10.1145/3544548.3580719

Hope, A., Schwaba, T., & Piper, A. M. (2014). Understanding digital and material social communications for older adults. In *Proceedings of the SIGCHI conference on human factors in computing systems* (pp. 3903–3912). https://doi.org/10.1145/2556288.2557133

Hornung, D., Müller, C., Shklovski, I., Jakobi, T., & Wulf, V. (2017). Navigating relationships and boundaries: Concerns around ICT-uptake for elderly people. In *Proceedings of the 2017 CHI conference on human factors in computing systems* (pp. 7057–7069). https://doi.org/10.1145/3025453.3025859

Jin, X., Tong, W., Wei, X., Wang, X., Kuang, E., Mo, X., Qu, H., & Fan, M. (2024). Exploring the opportunity of augmented reality (AR) in supporting older adults to explore and learn smartphone applications. In *Proceedings of the CHI conference on human factors in computing systems* (pp. 1–18). https://doi.org/10.1145/3613904.3641901

Joshi, K. R., Ulabhaje, A. A., Nataraj, R., & Martin-Hammond, A. (2025). *Skeptical yet curious: Attitudes towards conversational assistants for supporting social engagement among older adults living alone.*

Kachouie, R., Sedighadeli, S., Khosla, R., & Chu, M.-T. (2014). Socially assistive robots in elderly care: A mixed-method systematic literature review. *International Journal of Human-Computer Interaction, 30*(5), 369–393. https://doi.org/10.1080/10447318.2013.873278

Khosravi, P., Rezvani, A., & Wiewiora, A. (2016). The impact of technology on older adults' social isolation. *Computers in Human Behavior, 63*, 594–603. https://doi.org/10.1016/j.chb.2016.05.092

Knowles, B., & Hanson, V. L. (2018a). Older adults' deployment of 'distrust.' *ACM Transactions on Computer-Human Interaction, 25*(4), 1–25. https://doi.org/10.1145/3196490

Knowles, B., & Hanson, V. L. (2018b). The wisdom of older technology (non)users. *Communications of the ACM, 61*(3), 72–77. https://doi.org/10.1145/3179995

Knowles, B., Hanson, V. L., Rogers, Y., Piper, A. M., Waycott, J., Davies, N., Ambe, A. H., Brewer, R. N., Chattopadhyay, D., Dee, M., Frohlich, D., Gutierrez-Lopez, M., Jelen, B., Lazar, A.,

Nielek, R., Pena, B. B., Roper, A., Schlager, M., Schulte, B., & Yuan, I. Y. (2021). The harm in conflating aging with accessibility. *Communications of the ACM, 64*(7), 66–71. https://doi.org/10.1145/3431280

Kraut, R., & Burke, M. (2015). Internet use and psychological well-being: EDects of activity and audience. *Communications of the ACM, 58*(12), 5425.

Kugler, L. (2023). Virtual influencers in the real world. *Communications of the ACM, 66*(3), 23–25. https://doi.org/10.1145/3579635

Latulipe, C., Dsouza, R., & Cumbers, M. (2022). Unofficial proxies: How close others help older adults with banking. In *CHI conference on human factors in computing systems* (pp. 1–13). https://doi.org/10.1145/3491102.3501845

Lee, C., & Coughlin, J. F. (2015). PERSPECTIVE: Older adults' adoption of technology: An integrated approach to identifying determinants and barriers. *Journal of Product Innovation Management, 32*(5), 747–759. https://doi.org/10.1111/jpim.12176

Lewis, S. L., & Maslin, M. A. (2015). Defining the anthropocene. *Nature, 519*(7542), 171–180. https://doi.org/10.1038/nature14258

Lifshitz, R., Nimrod, G., & Bachner, Y. G. (2018). Internet use and well-being in later life: A functional approach. *Aging and Mental Health, 22*(1), 85–91. https://doi.org/10.1080/13607863.2016.1232370

Liu, Y., Goncalves, J., Ferreira, D., Xiao, B., Hosio, S., & Kostakos, V. (2014). CHI 1994–2013: Mapping two decades of intellectual progress through co-word analysis. In *Proceedings of the SIGCHI conference on human factors in computing systems* (pp. 3553–3562). https://doi.org/10.1145/2556288.2556969

Liu, C., Su, M., Xiang, Y., Huang, Y., Yang, Y., Zhang, K., & Fan, M. (2025). Toward enabling natural conversation with older adults via the design of LLM-powered voice agents that support interruptions and backchannels. In *Proceedings of the 2025 CHI conference on human factors in computing systems* (pp. 1–22). https://doi.org/10.1145/3706598.3714228

Liu, L., Stroulia, E., Nikolaidis, I., Miguel-Cruz, A., & Rios Rincon, A. (2016). Smart homes and home health monitoring technologies for older adults: A systematic review. *International Journal of Medical Informatics, 91*, 44–59. https://doi.org/10.1016/j.ijmedinf.2016.04.007

Lupton, D. (2024). Towards a gerontology of everything: A more-than-human perspective. *Journal of Aging Studies, 71*, 101278. https://doi.org/10.1016/j.jaging.2024.101278

Macedo, I. M. (2017). Predicting the acceptance and use of information and communication technology by older adults: An empirical examination of the revised UTAUT2. *Computers in Human Behavior, 75*, 935–948. https://doi.org/10.1016/j.chb.2017.06.013

Mack, K., McDonnell, E., Jain, D., Lu Wang, L., E. Froehlich, J., & Findlater, L. (2021). What do we mean by "accessibility research"? A literature survey of accessibility papers in CHI and ASSETS from 1994 to 2019. In *Proceedings of the 2021 CHI conference on human factors in computing systems* (pp. 1–18). https://doi.org/10.1145/3411764.3445412

Masoli, J. A. H., Todd, O., Burton, J. K., Wol, C., Walesby, K. E., Hewitt, J., Conroy, S., Van Oppen, J., Wilkinson, C., Evans, R., Anand, A., Hollinghurst, J., Bhanu, C., Keevil, V. L., Vardy, E. R. L., et al. (2023). New horizons in the role of digital data in the healthcare of older people. *Age and Ageing, 52*(8), afad134.

Montuwy, A., Cahour, B., & Dommes, A. (2018). Older pedestrians navigating with AR glasses and bone conduction headset. In *Extended abstracts of the 2018 CHI conference on human factors in computing systems* (pp. 1–6). https://doi.org/10.1145/3170427.3188503

Norman, D. (2023). *Design for a better world*.

Ogonowski, C., Aal, K., Vaziri, D., Rekowski, T. V., Randall, D., Schreiber, D., Wieching, R., & Wulf, V. (2016). ICT-based fall prevention system for older adults: Qualitative results from a long-term field study. *ACM Transactions on Computer–human Interaction, 23*(5), 1–33. https://doi.org/10.1145/2967102

Park, J. S., Bernstein, M. S., Brewer, R. N., Kamar, E., & Morris, M. R. (2021). Understanding the representation and representativeness of age in AI data sets. In *Proceedings of the 2021 AAAI/ACM conference on AI, ethics, and society* (pp. 834–842). https://doi.org/10.1145/3461702.3462590

Parra, C., Silveira, P., Far, I. K., Daniel, F., De Bruin, E. D., Cernuzzi, L., D'Andrea, V., & Casati, F. (2014). Information technology for active ageing: A review of theory and practice. *Foundations and Trends in Human–computer Interaction, 7*(4), 351–448.

Peine, A., & Neven, L. (2019). From intervention to co-constitution: New directions in theorizing about aging and technology. *The Gerontologist, 59*(1), 15–21. https://doi.org/10.1093/geront/gny050

Pradhan, A., Chopra, S., Upadhyay, P., Brewer, R., & Lazar, A. (2025). Towards accounting for non-human agency in technology design for aging. In *Proceedings of the 18th ACM international conference on PErvasive technologies related to assistive environments* (pp. 273–276). https://doi.org/10.1145/3733155.3733205

Rasi-Heikkinen, P., Rivinen, S., & Ahtinen, A. (2024). Older adults and robot literacy. *Educational Gerontology, 15*, 1–11. https://doi.org/10.1080/03601277.2024.2412367

Rogers, W. A., & Mitzner, T. L. (2017). Envisioning the future for older adults: Autonomy, health, well-being, and social connectedness with technology support. *Futures, 87*, 133–139. https://doi.org/10.1016/j.futures.2016.07.002

Rosales, A., & Fernández-Ardèvol, M. (2019). Structural ageism in big data approaches. *Nordicom Review, 40*, 51–64. https://doi.org/10.2478/nor-2019-0013

Sánchez-Chamorro, L., Toebosch, R., & Lallemand, C. (2024). Manipulative design and older adults: Co-creating magic machines to understand experiences of online manipulation. *Designing Interactive Systems Conference, 24*, 668–684. https://doi.org/10.1145/3643834.3661513

Sayago, S. (2023). *Cultures in human–computer interaction.* Springer.

Sayago, S., Forbes, P., & Blat, J. (2013). Older people becoming successful ICT learners over time: Challenges and strategies through an ethnographical lens. *Educational Gerontology, 39*(7), 527–544.

Sayago, S., Sloan, D., & Blat, J. (2011). Everyday use of computer-mediated communication tools and its evolution over time: An ethnographical study with older people. *Interacting with Computers, 23*(5), 543–554.

Sharma, V., Kumar, N., & Nardi, B. (2024). Post-growth human–computer interaction. *ACM Transactions on Computer-Human Interaction, 31*(1), 1–37. https://doi.org/10.1145/3624981

Sheahan, J., Cardamone, E., & Vines, J. (2024). Exploring generative postcard futures with older adults. In *Adjunct proceedings of the 2024 Nordic conference on human–computer interaction* (pp. 1–6). https://doi.org/10.1145/3677045.3685502

Shneiderman, B. (2022). *Human-centered AI.* Oxford University Press.

Tanprasert, T., Dai, J., & McGrenere, J. (2024). HelpCall: Designing informal technology assistance for older adults via videoconferencing. In *Proceedings of the CHI conference on human factors in computing systems* (pp. 1–23). https://doi.org/10.1145/3613904.3642938

Thach, K. S., Lederman, R., & Waycott, J. (2020). How older adults respond to the use of virtual reality for enrichment: A systematic review. In *Proceedings of the 32nd Australian conference on human–computer interaction* (pp. 303–313). https://doi.org/10.1145/3441000.3441003

Torgersson, O., Bekker, T., Barendregt, W., Eriksson, E., & Frauenberger, C. (2019). Making the child-computer interaction field grow up. *Interactions, 26*(2), 7–8. https://doi.org/10.1145/3310253

Verbeek, P.-P. (2015). Beyond interaction: A short introduction to mediation theory. *Interactions, 3*, 4565.

Wiberg, M. (2026). *Fast and slow.* CRC Press, Taylor & Francis Group.

Wilson, S. A., Byrne, P., Rodgers, S. E., & Maden, M. (2022). A systematic review of smartphone and tablet use by older adults with and without cognitive impairment. *Innovation in Aging, 6*(2), igac002.

Xie, B., Huang, M., & Watkins, I. (2012). *Technology and retirement life: A systematic review of the literature on older adults and social media.* Oxford University Press.

Xu, W., & Gao, Z. (2024). Applying HCAI in developing effective human-AI teaming: A perspective from human-AI joint cognitive systems. *Interactions, 31*, 32–37. https://doi.org/10.1145/363 5116

Yusif, S., Soar, J., & Hafeez-Baig, A. (2016). Older people, assistive technologies, and the barriers to adoption: A systematic review. *International Journal of Medical Informatics, 94*, 112–116. https://doi.org/10.1016/j.ijmedinf.2016.07.004

Zhai, Y., Xue, X., Guo, Z., Jin, T., Diao, Y., & Jeung, J. (2025). Hear us, then protect us: Navigating deepfake scams and safeguard interventions with older adults through participatory design. In *Proceedings of the 2025 CHI conference on human factors in computing systems* (pp. 1–19). https://doi.org/10.1145/3706598.3714423

Zhang, W., Yin, J., Chan, K. I., Sun, T., Jin, T., Jeung, J., & Gong, J. (2024). Beyond digital privacy: Uncovering deeper attitudes toward privacy in cameras among older adults. *International Journal of Human-Computer Studies, 192*, 103345. https://doi.org/10.1016/j.ijhcs.2024.103345

Zhou, S., Lin, W., Xu, Z., Wei, X., Huang, R., Ma, X., & Fan, M. (2025). JournalAIde: Empowering older adults in digital journal writing. In *Proceedings of the 2025 CHI conference on human factors in computing systems* (pp. 1–19). https://doi.org/10.1145/3706598.3713339

Bibliography

6

References that are not cited in the chapters are listed here:

Aalbers, T., Baars, M. A. E., & Rikkert, M. G. M. O. (2011). Characteristics of effective internet-mediated interventions to change lifestyle in people aged 50 and older: A systematic review. *Ageing Research Reviews, 10*(4), 487–497. https://doi.org/10.1016/j.arr.2011.05.001

Abdi, S., Witte, L. D., & Hawley, M. (2021). Exploring the potential of emerging technologies to meet the care and support needs of older people: A Delphi survey. *Geriatrics, 6*(1), 19. https://doi.org/10.3390/geriatrics6010019

Aceros, J. C., Pols, J., & Domènech, M. (2015). Where is grandma? Home telecare, good aging and the domestication of later life. *Technological Forecasting and Social Change, 93*, 102–111. https://doi.org/10.1016/j.techfore.2014.01.016

Adjei, S., Sam, S., Sekyere, F., & Boateng, P. (2022). Sign language interpreter-mediated qualitative interview with deaf participants in Ghana: Some methodological reflections for practice. *The Qualitative Report.* https://doi.org/10.46743/2160-3715/2022.5087

Ahmad, B. (2020). *ReDEAP: Recommendations for Developing Smartphone Applications for an Ageing Population.*

Ajaykumar, G., Pineda, K. T., & Huang, C.-M. (2023). Older adults' expectations, experiences, and preferences in programming physical robot assistance. *International Journal of Human–Computer Studies, 180*, 103127. https://doi.org/10.1016/j.ijhcs.2023.103127

Alexandrakis, D., Chorianopoulos, K., & Tselios, N. (2020). Implicit factors related to Greek older adults' perceived usability of online technologies: An exploratory study.

In *Proceedings of the 13th ACM international conference on PErvasive technologies related to assistive environments* (pp. 1–4). https://doi.org/10.1145/3389189.3393741

Alhouli, S. Y., Almania, N. A., Ahmad, M., & Sahoo, D. (2025). Designing a multimodal robot pet for older adults by young adults. In *Proceedings of the extended abstracts of the CHI conference on human factors in computing systems* (pp. 1–9). https://doi.org/10.1145/3706599.3720083

Alsulami, M. H., & Atkins, A. S. (2016). Factors influencing ageing population for adopting ambient assisted living technologies in the Kingdom of Saudi Arabia. *Ageing International, 41*(3), 227–239. https://doi.org/10.1007/s12126-016-9246-6

Altmeyer, M., Lessel, P., & Krüger, A. (2018). Investigating gamification for seniors aged 75+. In *Proceedings of the 2018 designing interactive systems conference* (pp. 453–458). https://doi.org/10.1145/3196709.3196799

Ambe, A. H., Roomkham, S., Lovell, D., & Brereton, M. (2023). Rethinking the design of human-data interaction through a study of older adults' wellbeing. In *Proceedings of the 35th Australian computer–human interaction conference* (pp. 266–279). https://doi.org/10.1145/3638380.3638451

Ambe, A. M. H. (2020). *From monitoring to engagement: Co-designing future technologies with older adults* [PhD, Queensland University of Technology]. https://doi.org/10.5204/thesis.eprints.204265

An, J., Zhu, X., Wan, K., Xiang, Z., Shi, Z., An, J., & Huang, W. (2024). Older adults' self-perception, technology anxiety, and intention to use digital public services. *BMC Public Health, 24*(1), 3533. https://doi.org/10.1186/s12889-024-21088-2

Anderson, B., & Tracey, K. (2001). Digital living: The impact (or otherwise) of the internet on everyday life. *American Behavioral Scientist, 45*(3), 456–475. https://doi.org/10.1177/00027640121957295

Angeli, A. D., Cozza, M., Jovanovic, M., Tonolli, L., Mushiba, M., McNeill, A., & Coventry, L. (n.d.). *Understanding motivations in designing for older adults.*

Arreola, I., Morris, Z., Francisco, M., Connelly, K., Caine, K., & White, G. (2014). From checking on to checking in: Designing for low socio-economic status older adults. In *Proceedings of the SIGCHI conference on human factors in computing systems* (pp. 1933–1936). https://doi.org/10.1145/2556288.2557084

Ashby, S., Hanna, J., Matos, S., Nash, C., & Faria, A. (2019). Fourth-wave HCI meets the twenty-first century manifesto. In *Proceedings of the halfway to the future symposium 2019* (pp. 1–11). https://doi.org/10.1145/3363384.3363467

Astell, A., Alm, N., Dye, R., Gowans, G., Vaughan, P., & Ellis, M. (2014). Digital video games for older adults with cognitive impairment. In K. Miesenberger, D. Fels, D. Archambault, P. Peňáz, & W. Zagler (Eds.), *Computers helping people with special needs* (Vol. 8547, pp. 264–271). Springer. https://doi.org/10.1007/978-3-319-08596-8_42

Aw, S., Koh, G. C., Oh, Y. J., Wong, M. L., Vrijhoef, H. J., Harding, S. C., Geronimo, M. A. B., & Hildon, Z. J. (2021). Interacting with place and mapping community needs to context: Comparing and triangulating multiple geospatial-qualitative methods using the focus–expand–compare approach. *Methodological Innovations*, *14*(1), 2059799120987772. https://doi.org/10.1177/2059799120987772

Baecker, R. M., Moffatt, K., & Massimi, M. (2012). *Technologies for aging gracefully*.

Bahadori, F., Abolfathi Momtaz, Y., Mohammadi Shahboulaghi, F., & Zandieh, Z. (2024). Information and communication technology adoption strategies among Iranian older adults: A qualitative evaluation. *Gerontology and Geriatric Medicine*, *10*, 23337214241246315. https://doi.org/10.1177/23337214241246315

Bailey, C., & Sheehan, C. (2009). Technology, older persons' perspectives and the anthropological ethnographic lens. *Alter*, *3*(2), 96–109. https://doi.org/10.1016/j.alter.2009.01.002

Bailey, C., Foran, T. G., Ni Scanaill, C., & Dromey, B. (2011). Older adults, falls and technologies for independent living: A life space approach. *Ageing and Society*, *31*(5), 829–848. https://doi.org/10.1017/S0144686X10001170

Baker, P. M. A., Bricout, J. C., Moon, N. W., Coughlan, B., & Pater, J. (2013). Communities of participation: A comparison of disability and aging identified groups on Facebook and LinkedIn. *Telematics and Informatics*, *30*(1), 22–34. https://doi.org/10.1016/j.tele.2012.03.004

Baker, S., Kelly, R. M., Waycott, J., Carrasco, R., Hoang, T., Batchelor, F., Ozanne, E., Dow, B., Warburton, J., & Vetere, F. (2019). Interrogating social virtual reality as a communication medium for older adults. *Proceedings of the ACM on Human–Computer Interaction*, *3*, 1–24. https://doi.org/10.1145/3359251

Baker, S., Waycott, J., Carrasco, R., Hoang, T., & Vetere, F. (2019). Exploring the design of social VR experiences with older adults. In *Proceedings of the 2019 on designing interactive systems conference* (pp. 303–315). https://doi.org/10.1145/3322276.3322361

Baker, S., Waycott, J., Carrasco, R., Kelly, R. M., Jones, A. J., Lilley, J., Dow, B., Batchelor, F., Hoang, T., & Vetere, F. (2021). Avatar-mediated communication in social VR: An in-depth exploration of older adult interaction in an emerging communication platform. In *Proceedings of the 2021 CHI conference on human factors in computing systems* (pp. 1–13). https://doi.org/10.1145/3411764.3445752

Baker, S., Waycott, J., Pedell, S., Hoang, T., & Ozanne, E. (2016). Older people and social participation: From touch-screens to virtual realities. In *Proceedings of the international symposium on interactive technology and ageing populations* (pp. 34–43). https://doi.org/10.1145/2996267.2996271

Ball, C., Francis, J., Huang, K.-T., Kadylak, T., Cotten, S. R., & Rikard, R. V. (2019). The physical–digital divide: Exploring the social gap between digital natives and physical natives. *Journal of Applied Gerontology*, *38*(8), 1167–1184. https://doi.org/10.1177/0733464817732518

Ballard, S., Chappell, K. M., & Kennedy, K. (2019). Judgment call the game: Using value sensitive design and design fiction to surface ethical concerns related to technology. In *Proceedings of the 2019 on designing interactive systems conference* (pp. 421–433). https://doi.org/10.1145/3322276.3323697

Baltes, P. B., & Smith, J. (2003). New frontiers in the future of aging: From successful aging of the young old to the dilemmas of the fourth age. *Gerontology, 49*(2), 123–135. https://doi.org/10.1159/000067946

Baltes, P. B., Mayer, K. U., & Berlin-Brandenburgische Akademie der Wissenschaften (Eds.). (1999). *The Berlin aging study: Aging from 70 to 100.* Cambridge University Press.

Baltes, P. B., Mayer, K. U., & Berlin-Brandenburgische Akademie der Wissenschaften (Eds.). (2001). *The Berlin aging study: Aging from 70 to 100; a research project of the Berlin-Brandenburg Academy of Sciences* (1st paperback print). Cambridge Univ. Press.

Barnard, Y., Bradley, M. D., Hodgson, F., & Lloyd, A. D. (2013). Learning to use new technologies by older adults: Perceived difficulties, experimentation behaviour and usability. *Computers in Human Behavior, 29*(4), 1715–1724. https://doi.org/10.1016/j.chb.2013.02.006

Barrantes Cáceres, R., & Cozzubo Chaparro, A. (2019). Age for learning, age for teaching: The role of inter-generational, intra-household learning in Internet use by older adults in Latin America. *Information, Communication and Society, 22*(2), 250–266. https://doi.org/10.1080/1369118X.2017.1371785

Batbold, T., Soro, A., & Schroeter, R. (2025). Driven by learning: Exploring older adults' pedagogical preferences in highly automated vehicle interactions. In *Proceedings of the extended abstracts of the CHI conference on human factors in computing systems* (pp. 1–7). https://doi.org/10.1145/3706599.3719842

Beh, J. (2019). *Interest in the learning of mobile touch screen technologies by older adults.*

Beh, J., Pedell, S., & Doube, W. (2015). Where is the 'I' in iPad? The role of interest in older adults' learning of mobile touch screen technologies. In *Proceedings of the annual meeting of the Australian special interest group for computer human interaction* (pp. 437–445). https://doi.org/10.1145/2838739.2838776

Behuniak, S. M. (2011). The living dead? The construction of people with Alzheimer's disease as zombies. *Ageing and Society, 31*(1), 70–92. https://doi.org/10.1017/S01446 86X10000693

Bell, C., Fausset, C., Farmer, S., Nguyen, J., Harley, L., & Fain, W. B. (2013). Examining social media use among older adults. In *Proceedings of the 24th ACM conference on hypertext and social media* (pp. 158–163). https://doi.org/10.1145/2481492.2481509

Benjamin-Thomas, T. E., Corrado, A. M., McGrath, C., Rudman, D. L., & Hand, C. (2018). Working towards the promise of participatory action research: Learning from ageing research exemplars. *International Journal of Qualitative Methods, 17*(1), 1609406918817953. https://doi.org/10.1177/1609406918817953

Bennett, B. (2019). Technology, ageing and human rights: Challenges for an ageing world. *International Journal of Law and Psychiatry, 66*, 101449. https://doi.org/10.1016/j.ijlp.2019.101449

Betts, L. R., Hill, R., & Gardner, S. E. (2019). "There's not enough knowledge out there": Examining older adults' perceptions of digital technology use and digital inclusion classes. *Journal of Applied Gerontology, 38*(8), 1147–1166. https://doi.org/10.1177/0733464817737621

Bhachu, A. S., Hine, N., & Arnott, J. (2008). Technology devices for older adults to aid self management of chronic health conditions. In *Proceedings of the 10th international ACM SIGACCESS conference on computers and accessibility* (pp. 59–66). https://doi.org/10.1145/1414471.1414484

Bickmore, T. W., Caruso, L., Clough-Gorr, K., & Heeren, T. (2005). 'It's just like you talk to a friend' relational agents for older adults. *Interacting with Computers, 17*(6), 711–735. https://doi.org/10.1016/j.intcom.2005.09.002

Bilius, L.-B., & Vatavu, R.-D. (2023). "I could wear it all of the time, just like my wedding ring:" Insights into older people's perceptions of smart rings. In *Extended abstracts of the 2023 CHI conference on human factors in computing systems* (pp. 1–8). https://doi.org/10.1145/3544549.3585771

Bilius, L.-B., Schipor, O. A., & Vatavu, R.-D. (2024). The age-reward perspective: A systematic review of reward mechanisms in serious games for older people. In *ACM international conference on interactive media experiences* (pp. 168–181). https://doi.org/10.1145/3639701.3656327

Binstock, R. H., & George, L. K. (2011). *Handbook of aging and the social sciences* (7th ed.). Elsevier/Academic Press.

Binstock, R. H., George, L. K., Cutler, S. J., Hendricks, J., & Schulz, J. H. (Eds.). (2006). *Handbook of aging and the social sciences* (6th ed.). Academic Press, an imprint of Elsevier.

Blackler, A., Mahar, D., & Popovic, V. (2010). Older adults, interface experience and cognitive decline. In *Proceedings of the 22nd conference of the computer–human interaction special interest group of Australia on computer–human interaction* (pp. 172–175). https://doi.org/10.1145/1952222.1952257

Blit-Cohen, E., & Litwin, H. (2004). Elder participation in cyberspace: A qualitative analysis of Israeli retirees. *Journal of Aging Studies, 18*(4), 385–398. https://doi.org/10.1016/j.jaging.2004.06.007

Bloch, N., & Bruce, B. C. (2011). Older adults and the new public sphere. In *Proceedings of the 2011 iConference* (pp. 1–7). https://doi.org/10.1145/1940761.1940762

Blouin Gagnon, L., Turcotte, S., DeBroux-Leduc, R., Gauthier, P., & Bier, N. (2025). Methods for co-designing health innovations with older adults: A rapid review. *Australasian Journal on Ageing, 44*(3), e70065. https://doi.org/10.1111/ajag.70065

Blythe, M., Steane, J., Roe, J., & Oliver, C. (2015). Solutionism, the game: Design fictions for positive aging. In *Proceedings of the 33rd annual ACM conference on human factors in computing systems* (pp. 3849–3858). https://doi.org/10.1145/2702123.2702491

Blythe, M., Wright, P., Bowers, J., Boucher, A., Jarvis, N., Reynolds, P., & Gaver, B. (2010). Age and experience: Ludic engagement in a residential care setting. In *Proceedings of the 8th ACM conference on designing interactive systems* (pp. 161–170). https://doi.org/10.1145/1858171.1858200

Bobeth, J., Deutsch, S., Schmehl, S., & Tscheligi, M. (n.d.). *Facing the user heterogeneity when designing touch interfaces for older adults: A representative personas approach.*

Bokolo Jnr, A. (2024). Examining the use of intelligent conversational voice-assistants for improved mobility behavior of older adults in smart cities. *International Journal of Human–Computer Interaction, 21*, 1–22. https://doi.org/10.1080/10447318.2024.2344145

Boot, W. R., Charness, N. H., Czaja, S. J., & Rogers, W. A. (2021). *Designing for older adults: Case studies, methods, and tools* (First edition). CRC Press, Taylor & Francis Group.

Boudiny, K. (2013). 'Active ageing': From empty rhetoric to effective policy tool. *Ageing and Society, 33*(6), 1077–1098. https://doi.org/10.1017/S0144686X1200030X

Bouma, H., Czaja, S. J., Umemuro, H., Rogers, W. A., Schulz, R., & Kurniawan, S. H. (2004). Technology: A means for enhancing the independence and connectivity of older people. In *CHI'04 extended abstracts on human factors in computing systems* (pp. 1580–1581). https://doi.org/10.1145/985921.986154

Bowen, S., Dearden, A., Wright, P., Wolstenholme, D., & Cobb, M. (2010). Participatory healthcare service design and innovation. In *Proceedings of the 11th biennial participatory design conference* (pp. 155–158). https://doi.org/10.1145/1900441.1900464

Branco, G., & Quaresma, M. (2024). *Towards active aging: Investigating innovations within intelligent communities.* DRS2024. https://doi.org/10.21606/drs.2024.1164

Brewer, R. N. (2022). "If Alexa knew the state I was in, it would cry": Older adults' perspectives of voice assistants for health. In *CHI conference on human factors in computing systems extended abstracts* (pp. 1–8). https://doi.org/10.1145/3491101.3519642

Brewer, R. N., & Piper, A. M. (2017). *xPress: Rethinking design for aging and accessibility through a voice-based online blogging community.*

Briede-Westermeyer, J. C., Pacheco-Blanco, B., Luzardo-Briceño, M., & Pérez-Villalobos, C. (2020). Mobile phone use by the elderly: Relationship between usability, social activity, and the environment. *Sustainability, 12*(7), 2690. https://doi.org/10.3390/su12072690

Briggs, P., & Thomas, L. (2015). An inclusive, value sensitive design perspective on future identity technologies. *ACM Transactions on Computer–Human Interaction, 22*(5), 1–28. https://doi.org/10.1145/2778972

Buch, E. D. (2015). Anthropology of aging and care. *Annual Review of Anthropology, 44*(1), 277–293. https://doi.org/10.1146/annurev-anthro-102214-014254

Burrows, A., Mitchell, V., & Nicolle, C. (2015). Cultural probes and levels of creativity. In *Proceedings of the 17th international conference on human–computer interaction with mobile devices and services adjunct* (pp. 920–923). https://doi.org/10.1145/2786567.2794302

Busch, P. A., Hausvik, G. I., Ropstad, O. K., & Pettersen, D. (2021). Smartphone usage among older adults. *Computers in Human Behavior, 121*, 106783. https://doi.org/10.1016/j.chb.2021.106783

Buse, C. E. (2009). When you retire, does everything become leisure? Information and communication technology use and the work/leisure boundary in retirement. *New Media and Society, 11*(7), 1143–1161. https://doi.org/10.1177/1461444809342052

Byun, C., Vasicek, P., & Seppi, K. (2023). Dispensing with humans in human–computer interaction research. In *Extended abstracts of the 2023 CHI conference on human factors in computing systems* (pp. 1–26). https://doi.org/10.1145/3544549.3582749

Cajamarca, G., Herskovic, V., Dondighual, S., Fuentes, C., & Verdezoto, N. (2023). Understanding how to design health data visualizations for Chilean older adults on mobile devices. In *Proceedings of the 2023 ACM designing interactive systems conference* (pp. 1309–1324). https://doi.org/10.1145/3563657.3596109

Caldeira, C., Bietz, M., Vidauri, M., & Chen, Y. (2017). Senior care for aging in place: Balancing assistance and independence. In *Proceedings of the 2017 ACM conference on computer supported cooperative work and social computing* (pp. 1605–1617). https://doi.org/10.1145/2998181.2998206

Caldeira, C., Nurain, N., Heintzman, A. A., Molchan, H., Caine, K., Demiris, G., Siek, K. A., Reeder, B., & Connelly, K. (2023). How do I compare to the other people? Older adults' perspectives on personal smart home data for self-management. In *Proceedings of the ACM on human–computer interaction* (pp. 1–32). https://doi.org/10.1145/3610029

Camargo, J. D., Silva, T., & Abreu, J. (2024). Always together: Combining TV notifications and voice interactions to connect older adults to other generations. In *ACM international conference on interactive media experiences* (pp. 388–393). https://doi.org/10.1145/3639701.3663642

Carpenter, B. D., & Buday, S. (2007). Computer use among older adults in a naturally occurring retirement community. *Computers in Human Behavior, 23*(6), 3012–3024. https://doi.org/10.1016/j.chb.2006.08.015

Carros, F., Meurer, J., Löffler, D., Unbehaun, D., Matthies, S., Koch, I., Wieching, R., Randall, D., Hassenzahl, M., & Wulf, V. (2020). Exploring human–robot interaction with the elderly: Results from a ten-week case study in a care home. In *Proceedings of the 2020 CHI conference on human factors in computing systems* (pp. 1–12). https://doi.org/10.1145/3313831.3376402

Castleton, A. (2019). *Ageing and technology: The case of a one-tablet-per-senior program in Uruguay* [Doctor of Philosophy, Carleton University]. https://doi.org/10.22215/etd/2019-13674

Castro-Rojas, M. D., Blanco-Molina, M., & Coto-Chotto, M. (2025). Information and communication technologies to promote healthy ageing: A rural community intervention during COVID-19 pandemic. *Human Technology, 21*(2), 374–394. https://doi.org/10.14254/1795-6889.2025.21-2.7

Cavender, A., Trewin, S., & Hanson, V. (2008). General writing guidelines for technology and people with disabilities. *ACM SIGACCESS Accessibility and Computing, 92*, 17–22. https://doi.org/10.1145/1452562.1452565

Cerna, K., & Müller, C. (2021). *Making online participatory design work: Understanding the digital ecologies of older adults.* https://doi.org/10.18420/ECSCW2021_N22

Chan, M. Y., Haber, S., Drew, L. M., & Park, D. C. (2016). Training older adults to use tablet computers: Does it enhance cognitive function? *The Gerontologist, 56*(3), 475–484. https://doi.org/10.1093/geront/gnu057

Charness, N., & Boot, W. R. (2009). Aging and information technology use: Potential and barriers. *Current Directions in Psychological Science, 18*(5), 253–258. https://doi.org/10.1111/j.1467-8721.2009.01647.x

Charness, N., & Boot, W. R. (2016). Technology, gaming, and social networking. In *Handbook of the psychology of aging* (pp. 389–407). Elsevier. https://doi.org/10.1016/B978-0-12-411469-2.00020-0

Charness, N., Dunlop, M., Munteanu, C., Nicol, E., Oulasvirta, A., Ren, X., Sarcar, S., & Silpasuwanchai, C. (2016). Rethinking mobile interfaces for older adults. In *Proceedings of the 2016 CHI conference extended abstracts on human factors in computing systems* (pp. 1131–1134). https://doi.org/10.1145/2851581.2886431

Chaudhry, B. M., Islam, M. U., & Chawla, N. V. (2024). Longitudinal evaluation of casual puzzle tablet games by older adults. *Designing Interactive Systems Conference, 12*, 2073–2087. https://doi.org/10.1145/3643834.3661528

Chen, K., & Chan, A. H. S. (2011). A review of technology acceptance by older adults. *Gerontechnology, 10*(1), 1–12. https://doi.org/10.4017/gt.2011.10.01.006.00

Chen, Y. (2009). *Usability analysis on online social networks for the elderly.*

Chen, Y., Liu, L., Wu, Q., & Yang, C.-F. (2025). Proactive and adaptive elderly-centered governance framework through synergistic integration of the internet of things and multi-agent systems. *Sensors and Materials, 37*(6), 2431. https://doi.org/10.18494/SAM5741

Chiu, C.-J., Hu, Y.-H., Lin, D.-C., Chang, F.-Y., Chang, C.-S., & Lai, C.-F. (2016). The attitudes, impact, and learning needs of older adults using apps on touchscreen mobile devices: Results from a pilot study. *Computers in Human Behavior, 63,* 189–197. https://doi.org/10.1016/j.chb.2016.05.020

Chou, W.-H., Lai, Y.-T., & Liu, K.-H. (2010). Decent digital social media for senior life: A practical design approach. In *Proceedings of the 2010 3rd international conference on computer science and information technology* (pp. 249–253). https://doi.org/10.1109/ICCSIT.2010.5565189

Choudrie, J., Junior, C.-O., McKenna, B., & Richter, S. (2018). Understanding and conceptualising the adoption, use and diffusion of mobile banking in older adults: A research agenda and conceptual framework. *Journal of Business Research, 88,* 449–465. https://doi.org/10.1016/j.jbusres.2017.11.029

Cid, A., Sotelo, R., Leguisamo, M., & Ramírez-Michelena, M. (2020). Tablets for deeply disadvantaged older adults: Challenges in long-term care facilities. *International Journal of Human–Computer Studies, 144,* 102504. https://doi.org/10.1016/j.ijhcs.2020.102504

Cifter, A. S., & Dong, H. (2010). Instruction manual usage: A comparison of younger people, older people and people with cognitive disabilities. In R. Winter, J. L. Zhao, & S. Aier (Eds.), *Global perspectives on design science research* (Vol. 6105, pp. 410–425). Springer. https://doi.org/10.1007/978-3-642-13335-0_28

Close, J. C. T. (2023). Big data—big opportunity. *Age and Ageing, 52*(1), afac262. https://doi.org/10.1093/ageing/afac262

Coelho, A. R. (2024). Modes of relating to the new ICTs among older internet users: A qualitative approach. *Ageing and Society, 44*(4), 963–987. https://doi.org/10.1017/S0144686X22000605

Coelho, J., & Duarte, C. (2016). A literature survey on older adults' use of social network services and social applications. *Computers in Human Behavior, 58,* 187–205. https://doi.org/10.1016/j.chb.2015.12.053

Coelho, J., Rito, F., Luz, N., & Duarte, C. (2015). Prototyping TV and tablet Facebook interfaces for older adults. In J. Abascal, S. Barbosa, M. Fetter, T. Gross, P. Palanque, & M. Winckler (Eds.), *Human–computer interaction: INTERACT 2015* (Vol. 9296, pp. 110–128). Springer. https://doi.org/10.1007/978-3-319-22701-6_9

Coghlan, S., Waycott, J., Lazar, A., & Barbosa Neves, B. (2021). Dignity, autonomy, and style of company: Dimensions older adults consider for robot companions. In *Proceedings of the ACM on human–computer interaction* (pp. 1–25). https://doi.org/10.1145/3449178

Colombo-Ruano, L., Rodríguez-Silva, C., Violant-Holz, V., & González-González, C. S. (2021). Technological acceptance of voice assistants in older adults: An online co-creation experience. In *Proceedings of the XXI international conference on human computer interaction* (pp. 1–5). https://doi.org/10.1145/3471391.3471432

Comerford, S., O'Kane, E., Roe, D., Alsharedah, H., O'Neill, B., Walsh, M., & Briggs, R. (2024). Everyday ageism experienced by community-dwelling older people with

frailty. *European Geriatric Medicine, 15*(6), 1763–1769. https://doi.org/10.1007/s41
999-024-01048-0

Compagna, D., & Kohlbacher, F. (2015). The limits of participatory technology
development: The case of service robots in care facilities for older people. *Technological Forecasting and Social Change, 93*, 19–31. https://doi.org/10.1016/j.techfore.
2014.07.012

Conci, M., Pianesi, F., & Zancanaro, M. (2009). Useful, social and enjoyable: Mobile
phone adoption by older people. In T. Gross, J. Gulliksen, P. Kotzé, L. Oestreicher, P.
Palanque, R. O. Prates, & M. Winckler (Eds.), *Human–computer interaction: INTER-
ACT 2009* (Vol. 5726, pp. 63–76). Springer. https://doi.org/10.1007/978-3-642-036
55-2_7

Conde, M., Mikhailova, V., & Döring, N. (2024). "I have the feeling that the person
is here": Older adults' attitudes, usage intentions, and requirements for a telepresence
robot. *International Journal of Social Robotics, 16*(7), 1619–1639. https://doi.org/10.
1007/s12369-024-01143-z

Connelly, K., Ur Rehman Laghari, K., Mokhtari, M., & Falk, T. H. (2014). Approaches
to understanding the impact of technologies for aging in place: A mini-review.
Gerontology, 60(3), 282–288. https://doi.org/10.1159/000355644

Cooper, S., Di Fava, A., Villacanas, O., Silva, T., Fernandez-Carbajales, V., Unzueta,
L., Serras, M., Marchionni, L., & Ferro, F. (2021). Social robotic application to support active and healthy ageing. In *Proceedings of the 2021 30th IEEE international
conference on robot and human interactive communication (RO-MAN)* (pp. 1074–1080).
https://doi.org/10.1109/RO-MAN50785.2021.9515432

Cornwell, B., Laumann, E. O., & Schumm, L. P. (2008). The social connectedness of
older adults: A national profile. *American Sociological Review, 73*(2), 185–203. https://
doi.org/10.1177/000312240807300201

Correia, C., Oliveira, E., & Nunes, F. (2020). Using illustration to create more inclusive
user interfaces for older adults. *Interactions, 27*(2), 79–81. https://doi.org/10.1145/337
8556

Cotten, S. R., Ghaiumy Anaraky, R., & Schuster, A. M. (2023). Social media use may
not be as bad as some suggest: Implication for older adults. *Innovation in Aging, 7*(3),
igad022. https://doi.org/10.1093/geroni/igad022

Cottle, K. E. (2017). *Current patterns of ownership and usage of mobile technology in
older adults.*

Coughlin, J. F. (2007). *New expectations from older users: Five lessons for product
design and innovation in an aging marketplace.*

Cresci, M. K., Jarosz, P. A., & Templin, T. N. (2012). Are health answers online
for older adults? *Educational Gerontology, 38*(1), 10–19. https://doi.org/10.1080/036
01277.2010.515890

Cuadra, A., Bethune, J., Krell, R., Lempel, A., Hänsel, K., Shahrokni, A., Estrin, D., &
Dell, N. (2023). Designing voice-first ambient interfaces to support aging in place. In

Proceedings of the 2023 ACM designing interactive systems conference (pp. 2189–2205). https://doi.org/10.1145/3563657.3596104

Culén, A. L., & Bratteteig, T. (n.d.). *Touch interfaces for elderly: Some design challenges.*

Curran, K., Walters, N., & Robinson, D. (2007). Investigating the problems faced by older adults and people with disabilities in online environments. *Behaviour and Information Technology, 26*(6), 447–453. https://doi.org/10.1080/01449290600740868

Czaja, S. J., Charness, N., Fisk, A. D., Hertzog, C., Nair, S. N., Rogers, W. A., & Sharit, J. (2006). Factors predicting the use of technology: Findings from the center for research and education on aging and technology enhancement (create). *Psychology and Aging, 21*(2), 333–352. https://doi.org/10.1037/0882-7974.21.2.333

D'Haeseleer, I., Gerling, K., Schreurs, D., Vanrumste, B., & Vanden Abeele, V. (2019). Ageing is not a disease: Pitfalls for the acceptance of self-management health systems supporting healthy ageing. In *The 21st international ACM SIGACCESS conference on computers and accessibility* (pp. 286–298). https://doi.org/10.1145/3308561.3353794

Dalgaard, L. G., Gronvall, E., & Verdezoto, N. (2013). MediFrame: A tablet application to plan, inform, remind and sustain older adults' medication intake. In *Proceedings of the 2013 IEEE international conference on healthcare informatics* (pp. 36–45). https://doi.org/10.1109/ICHI.2013.12

Dare, J., Wilkinson, C., Donovan, R., Lo, J., McDermott, M.-L., O'Sullivan, H., & Marquis, R. (2019). Guidance for research on social isolation, loneliness, and participation among older people: Lessons from a mixed methods study. *International Journal of Qualitative Methods, 18*, 1609406919872914. https://doi.org/10.1177/1609406919872914

Darroch, I., Goodman, J., Brewster, S., & Gray, P. (2005). The effect of age and font size on reading text on handheld computers. In M. F. Costabile & F. Paternò (Eds.), *Human–computer interaction—INTERACT 2005* (Vol. 3585, pp. 253–266). Springer. https://doi.org/10.1007/11555261_23

Davidson, J. L., Naik, R., Mannan, U. A., Azarbakht, A., & Jensen, C. (2014). On older adults in free/open source software: Reflections of contributors and community leaders. In *Proceedings of the 2014 IEEE symposium on visual languages and human-centric computing (VL/HCC)* (pp. 93–100). https://doi.org/10.1109/VLHCC.2014.6883029

Davis, F. D., & Venkatesh, V. (2004). Toward preprototype user acceptance testing of new information systems: Implications for software project management. *IEEE Transactions on Engineering Management, 51*(1), 31–46. https://doi.org/10.1109/TEM.2003.822468

De Angeli, A., Jovanović, M., McNeill, A., & Coventry, L. (2020). Desires for active ageing technology. *International Journal of Human–Computer Studies, 138*, 102412. https://doi.org/10.1016/j.ijhcs.2020.102412

De Barros, A. C., Rêgo, S., & Antunes, J. (2014). Aspects of human-centred design in hci with older adults: Experiences from the field. In S. Sauer, C. Bogdan, P. Forbrig, R. Bernhaupt, & M. Winckler (Eds.), *Human-centered software engineering* (Vol. 8742, pp. 235–242). Springer. https://doi.org/10.1007/978-3-662-44811-3_14

De Lara, S. M. A., Fortes, R. P. D. M., Russo, C. M., & Freire, A. P. (2016). A study on the acceptance of website interaction aids by older adults. *Universal Access in the Information Society, 15*(3), 445–460. https://doi.org/10.1007/s10209-015-0419-y

De Schutter, B., & Vanden Abeele, V. (2015). Towards a gerontoludic manifesto. *Anthropology and Aging, 36*(2), 112–120. https://doi.org/10.5195/aa.2015.104

Delft University of Technology, Giaccardia, E., Kuijerb, L., & Nevenc, L. (2016). Design for resourceful ageing: Intervening in the ethics of gerontechnology. In *Design research society conference 2016*. https://doi.org/10.21606/drs.2016.258

Delgado, F., Yang, S., Madaio, M., & Yang, Q. (2023). The participatory turn in AI design: Theoretical foundations and the current state of practice. *Equity and Access in Algorithms, Mechanisms, and Optimization, 94*, 1–23. https://doi.org/10.1145/3617694.3623261

Designing User Interfaces for Older Adults: Myth Busters: UXmatters. (n.d.).

Dhillon, J. S., Wünsche, B. C., & Lutteroth, C. (2013). An online social-networking enabled telehealth system for seniors: A case study. *User Interfaces, 139*.

Diaz-Orueta, U., Hopper, L., & Konstantinidis, E. (2020). Shaping technologies for older adults with and without dementia: Reflections on ethics and preferences. *Health Informatics Journal, 26*(4), 3215–3230. https://doi.org/10.1177/1460458219899590

Diaz, M., Johnson, I., Lazar, A., Piper, A. M., & Gergle, D. (2018). Addressing age-related bias in sentiment analysis. In *Proceedings of the 2018 CHI conference on human factors in computing systems* (pp. 1–14). https://doi.org/10.1145/3173574.3173986

Dickinson, A., Eisma, R., & Gregor, P. (2011). The barriers that older novices encounter to computer use. *Universal Access in the Information Society, 10*(3), 261–266. https://doi.org/10.1007/s10209-010-0208-6

Dienlin, T., & Johannes, N. (2020). The impact of digital technology use on adolescent well-being. *Dialogues in Clinical Neuroscience, 22*(2), 135–142. https://doi.org/10.31887/DCNS.2020.22.2/tdienlin

Dishman, E., & Scannell, N. (n.d.). *Community supports for ageing 3*.

Dogruel, L., Joeckel, S., & Bowman, N. D. (2013). Elderly people and morality in virtual worlds: A cross-cultural analysis of elderly people's morality in interactive media. *New Media and Society, 15*(2), 276–293. https://doi.org/10.1177/1461444812451571

Druin, A. (2002). The role of children in the design of new technology. *Behaviour and Information Technology, 21*(1), 1–25. https://doi.org/10.1080/01449290110108659

Duque, M., Pink, S., Strengers, Y., Martin, R., & Nicholls, L. (2021). Automation, wellbeing and digital voice assistants: Older people and Google devices. *Convergence: The International Journal of Research into New Media Technologies, 27*(5), 1189–1206. https://doi.org/10.1177/13548565211038537

Edwards, M. A., & Roy, S. (2017). Academic research in the twenty-first century: Maintaining scientific integrity in a climate of perverse incentives and hypercompetition. *Environmental Engineering Science, 34*(1), 51–61. https://doi.org/10.1089/ees. 2016.0223

Eisma, R. (2003). *Mutual inspiration in the development of new technology for older people.*

Eronen, J., Paakkari, L., Portegijs, E., Saajanaho, M., & Rantanen, T. (2019). Assessment of health literacy among older Finns. *Aging Clinical and Experimental Research, 31*(4), 549–556. https://doi.org/10.1007/s40520-018-1104-9

Esposito, A., Amorese, T., Cuciniello, M., Esposito, A. M., Troncone, A., Torres, M. I., Schlögl, S., & Cordasco, G. (2019). Seniors' acceptance of virtual humanoid agents. In A. Leone, A. Caroppo, G. Rescio, G. Diraco, & P. Siciliano (Eds.), *Ambient assisted living* (Vol. 544, pp. 429–443). Springer. https://doi.org/10.1007/978-3-030-05921-7_35

Etkin, K. (2022). *AgeTech revolution: A book about the intersection of aging and technology: A book about the intersection of aging and technology.* New Degree Press.

European Commission. Directorate General for the Information Society and Media. (2013). *Final evaluation of the ambient assisted living joint programme.* Publications Office. https://doi.org/10.2759/361

Ezer, N. (2008). *A dissertation presented to the academic faculty.*

Fan, C., Forlizzi, J., & Dey, A. (2012). Considerations for technology that support physical activity by older adults. In *Proceedings of the 14th international ACM SIGACCESS conference on computers and accessibility* (pp. 33–40). https://doi.org/10.1145/2384916.2384923

Fan, M., & Truong, K. N. (2018). Guidelines for creating senior-friendly product instructions. *ACM Transactions on Accessible Computing, 11*(2), 1–35. https://doi.org/10.1145/3209882

Fan, M., Zhao, Q., & Tibdewal, V. (2021). Older adults' think-aloud verbalizations and speech features for identifying user experience problems. In *Proceedings of the 2021 CHI conference on human factors in computing systems* (pp. 1–13). https://doi.org/10.1145/3411764.3445680

Farina, K., & Nitsche, M. (2015). Outside the brick: Exploring prototyping for the elderly. In *Proceedings of the 2015 British HCI conference* (pp. 11–17). https://doi.org/10.1145/2783446.2783571

Feist, H., Parker, K., Howard, N., & Hugo, G. (2010). *New technologies: Their potential role in linking rural older people to community, 8*(2).

Felsted, K. F., & Wright, S. D. (2014). *Toward post ageing: technology in an ageing society* (Vol. 1). Springer. https://doi.org/10.1007/978-3-319-09051-1

Findlater, L., Froehlich, J. E., Fattal, K., Wobbrock, J. O., & Dastyar, T. (2013). Age-related differences in performance with touchscreens compared to traditional mouse input. In *Proceedings of the SIGCHI conference on human factors in computing systems* (pp. 343–346). https://doi.org/10.1145/2470654.2470703

Finn, K., & Johnson, J. (2013). A usability study of websites for older travelers. In C. Stephanidis & M. Antona (Eds.), *Universal access in human–computer interaction. User and context diversity* (Vol. 8010, pp. 59–67). Springer. https://doi.org/10.1007/978-3-642-39191-0_7

Ford, J. H., Daley, R. T., & Kensinger, E. A. (2024). The benefits of socioemotional learning strategies and video formats for older digital immigrants learning a novel smartphone application. *Frontiers in Aging, 5*, 1416139. https://doi.org/10.3389/fragi.2024.1416139

Forsman, A. K., & Nordmyr, J. (2017). Psychosocial links between internet use and mental health in later life: A systematic review of quantitative and qualitative evidence. *Journal of Applied Gerontology, 36*(12), 1471–1518. https://doi.org/10.1177/0733464815595509

Foucault, M. (1982). *WhyStudyPower? The questionof the subject.*

Francese, R., & Risi, M. (2016). Supporting elderly people by ad hoc generated mobile applications based on vocal interaction. *Future Internet, 8*(3), 42. https://doi.org/10.3390/fi8030042

Franz, R. L., Findlater, L., Barbosa Neves, B., & Wobbrock, J. O. (2019). Gender and help seeking by older adults when learning new technologies. In *The 21st international ACM SIGACCESS conference on computers and accessibility* (pp. 136–142). https://doi.org/10.1145/3308561.3353807

Franz, R. L., Wobbrock, J. O., Cheng, Y., & Findlater, L. (2019). Perception and adoption of mobile accessibility features by older adults experiencing ability changes. In *The 21st international ACM SIGACCESS conference on computers and accessibility* (pp. 267–278). https://doi.org/10.1145/3308561.3353780

Friemel, T. N. (2016). The digital divide has grown old: Determinants of a digital divide among seniors. *New Media and Society, 18*(2), 313–331. https://doi.org/10.1177/1461444814538648

Frik, A., & Nurgalieva, L. (n.d.). *Privacy and security threat models and mitigation strategies of older adults.*

Frik, A., Nurgalieva, L., Bernd, J., Lee, J. S., Schaub, F., & Egelman, S. (2019). *Privacy and security threat models and mitigation strategies of older adults.*

Galindo Esparza, R. P., Dudley, J. J., Garaj, V., & Kristensson, P. O. (2025). Exclusion rates among disabled and older users of virtual and augmented reality. In *Proceedings of the 2025 CHI conference on human factors in computing systems* (pp. 1–15). https://doi.org/10.1145/3706598.3713377

Gallistl, V., Banday, M. U. L., Berridge, C., Grigorovich, A., Jarke, J., Mannheim, I., Marshall, B., Martin, W., Moreira, T., Van Leersum, C. M., & Peine, A. (2024). Addressing the black box of AI: A model and research agenda on the co-constitution of aging and artificial intelligence. *The Gerontologist, 64*(6), gnae039. https://doi.org/10.1093/geront/gnae039

Gallud, J. A., Fardoun, H. M., Andres, F., & Safa, N. (2018). A study on how older people use emojis. In *Proceedings of the XIX international conference on human computer interaction* (pp. 1–4). https://doi.org/10.1145/3233824.3233861

Gao, Q., & Zhou, J. (Eds.). (2024). *Human aspects of IT for the aged population: 10th international conference, ITAP 2024, held as part of the 26th HCI international conference, HCII 2024, Washington, DC, USA, June 29–July 4, 2024, proceedings, part I* (Vol. 14725). Springer. https://doi.org/10.1007/978-3-031-61543-6

García, A. A., Fuentes, J. V., & Rodriguez, R. P. (n.d.). *Un per l de las personas mayores en España, 2014.*

Gasteiger, N., Ahn, H. S., Fok, C., Lim, J., Lee, C., MacDonald, B. A., Kim, G. H., & Broadbent, E. (2022). Older adults' experiences and perceptions of living with Bomy, an assistive dailycare robot: A qualitative study. *Assistive Technology, 34*(4), 487–497. https://doi.org/10.1080/10400435.2021.1877210

Gaver, W., & Gaver, F. (2023). Living with light touch: An autoethnography of a simple communication device in long-term use. In *Proceedings of the 2023 CHI conference on human factors in computing systems* (pp. 1–14). https://doi.org/10.1145/3544548.3580807

Gee, N. R., Mueller, M. K., & Curl, A. L. (2017). Human–animal interaction and older adults: An overview. *Frontiers in Psychology, 8*, 1416. https://doi.org/10.3389/fpsyg.2017.01416

Gerling, K. M., Mandryk, R. L., & Linehan, C. (2015). Long-term use of motion-based video games in care home settings. In *Proceedings of the 33rd annual ACM conference on human factors in computing systems* (pp. 1573–1582). https://doi.org/10.1145/2702123.2702125

Gewirtz-Meydan, A., Opuda, E., & Ayalon, L. (2023). Sex and love among older adults in the digital world: A scoping review. *The Gerontologist, 63*(2), 218–230. https://doi.org/10.1093/geront/gnac093

Ghaiumy Anaraky, R., Schuster, A. M., & Cotten, S. R. (2024). Can changes in older adults' technology use patterns be used to detect cognitive decline? *The Gerontologist, 64*(7), gnad158. https://doi.org/10.1093/geront/gnad158

Giaccardi, E., Cila, N., Speed, C., & Caldwell, M. (2016). Thing ethnography: Doing design research with non-humans. In *Proceedings of the 2016 ACM conference on designing interactive systems* (pp. 377–387). https://doi.org/10.1145/2901790.2901905

Gilbertson, T. (2014). Attitudes and behaviours towards web accessibility and ageing: Results of an industry survey. *Gerontechnology, 13*(3), 337–344. https://doi.org/10.4017/gt.2015.13.3.004.00

Giles, H., & Gasiorek, J. (2011). Intergenerational communication practices. In *Handbook of the psychology of aging* (pp. 233–247). Elsevier. https://doi.org/10.1016/B978-0-12-380882-0.00015-2

Gilleard, C., & Higgs, P. (2002). The third age: Class, cohort or generation? *Ageing and Society, 22*(3), 369–382. https://doi.org/10.1017/S0144686X0200870X

Gilleard, C., & Higgs, P. (2005). *Contexts of ageing. Class, cohort and community.* Polity press.

Giusti, L., Mencarini, E., & Zancanaro, M. (2010). 'Luckily, I don't need it': Elderly and the use of artifacts for time management. In *Proceedings of the 6th Nordic conference on human–computer interaction: Extending boundaries* (pp. 198–206). https://doi.org/10.1145/1868914.1868940

Golant, S. M. (2017). A theoretical model to explain the smart technology adoption behaviors of elder consumers (Elderadopt). *Journal of Aging Studies, 42*, 56–73. https://doi.org/10.1016/j.jaging.2017.07.003

Gomez-Hernandez, M. (2024). Industry visions of technology for older adults: A futures anthropology perspective. *Journal of Aging Studies, 70*, 101248. https://doi.org/10.1016/j.jaging.2024.101248

González-González, C., Violant-Holz, V., Colombo-Ruano, L., & Rodríguez-Silva, C. (2025). Designing ethical and inclusive voice assistants through co-creation with older adults. *Behaviour and Information Technology, 251*, 1–23. https://doi.org/10.1080/0144929X.2025.2514245

Goodman, J., Dickinson, A., & Syme, A. (2004). Gathering requirements for mobile devices using focus groups with older people. In S. Keates, J. Clarkson, P. Langdon, & P. Robinson (Eds.), *Designing a more inclusive world* (pp. 81–90). Springer. https://doi.org/10.1007/978-0-85729-372-5_9

Graf, C., Hochleitner, C., & Tscheligi, M. (2012). *How to design accessible TVMs for older adults.*

Graham, L., Brundle, C., Harrison, N., Andre, D., Clegg, A., Forster, A., & Spilsbury, K. (2024). What are the priorities for research of older people living in their own home, including those living with frailty? A systematic review and content analysis of studies reporting older people's priorities and unmet needs. *Age and Ageing, 53*(1), afad232. https://doi.org/10.1093/ageing/afad232

Gramß, D., & Struve, D. (2009). Instructional videos for supporting older adults who use interactive systems. *Educational Gerontology, 35*(2), 164–176. https://doi.org/10.1080/03601270802421434

Greenhalgh, T., Procter, R., Wherton, J., Sugarhood, P., & Shaw, S. (2012). The organising vision for telehealth and telecare: Discourse analysis. *BMJ Open, 2*(4), e001574. https://doi.org/10.1136/bmjopen-2012-001574

Gregor, P., AppliedComputing, D., & Newell, A. F. (2001). *Designing for dynamic diversity m making accessible interfaces for older people.*

Grosinger, J., Vetere, F., & Fitzpatrick, G. (2012). Agile life: Addressing knowledge and social motivations for active aging. In *Proceedings of the 24th Australian computer–human interaction conference* (pp. 162–165). https://doi.org/10.1145/2414536.2414566

Gu, X., Hamido, S., & Itoh, K. (2024). Older adults' awareness, motivation, and behavior changes by wearable activity trackers before and during the COVID-19 pandemic. *Gerontechnology*, *23*(1), 1–13. https://doi.org/10.4017/gt.2024.23.1.842.02

Guess, A., Nagler, J., & Tucker, J. (2019). Less than you think: Prevalence and predictors of fake news dissemination on Facebook. *Science Advances*, *5*(1), eaau4586. https://doi.org/10.1126/sciadv.aau4586

Haan, M. D., Brankaert, R., Kenning, G., & Lu, Y. (2021). Creating a social learning environment for and by older adults in the use and adoption of smartphone technology to age in place. *Frontiers in Public Health*, *9*, 568822. https://doi.org/10.3389/fpubh.2021.568822

Hadi Mogavi, R., Son, J., Yang, S., Wang, D. M., Choong, L., Alhilal, A., Zhou, P. Y., Hui, P., & Nacke, L. E. (2024). The jade gateway to exergaming: How socio-cultural factors shape exergaming among East Asian older adults. *Proceedings of the ACM on Human–Computer Interaction*, *8*, 1–34. https://doi.org/10.1145/3677106

Haghzare, S., Campos, J., & Mihailidis, A. (2018). Identifying the factors influencing older adults' perceptions of fully automated vehicles. In *Adjunct proceedings of the 10th international conference on automotive user interfaces and interactive vehicular applications* (pp. 98–103). https://doi.org/10.1145/3239092.3265949

Hakkarainen, P. (2012). 'No good for shovelling snow and carrying firewood': Social representations of computers and the internet by elderly Finnish non-users. *New Media and Society*, *14*(7), 1198–1215. https://doi.org/10.1177/1461444812442663

Hall, S., Longhurst, S., & Higginson, I. J. (2009). Challenges to conducting research with older people living in nursing homes. *BMC Geriatrics*, *9*(1), 38. https://doi.org/10.1186/1471-2318-9-38

Hallewell Haslwanter, J. D., & Fitzpatrick, G. (2017). Why do few assistive technology systems make it to market? The case of the HandyHelper project. *Universal Access in the Information Society*, *16*(3), 755–773. https://doi.org/10.1007/s10209-016-0499-3

Hanson, M. A., Barreiro, P. G., Crosetto, P., & Brockington, D. (2024). The strain on scientific publishing. *Quantitative Science Studies*, *5*(4), 823–843. https://doi.org/10.1162/qss_a_00327

Hanson, V. L. (2009). Cognition, age, and web browsing. In C. Stephanidis (Ed.), *Universal access in human–computer interaction: Addressing diversity* (Vol. 5614, pp. 245–250). Springer. https://doi.org/10.1007/978-3-642-02707-9_27

Hanson, V. L. (2010). Influencing technology adoption by older adults. *Interacting with Computers*, *22*(6), 502–509. https://doi.org/10.1016/j.intcom.2010.09.001

Hanson, V. L., & Gibson, L. (2010). *Engaging those who are disinterested: Access for digitally excluded older adults.*

Hao, Y., Liu, Z., Riter, R. N., & Kalantari, S. (2024). Advancing patient-centered shared decision-making with AI systems for older adult cancer patients. In *Proceedings of the CHI conference on human factors in computing systems* (pp. 1–20). https://doi.org/10.1145/3613904.3642353

Hargittai, E., Piper, A. M., & Morris, M. R. (2019). From internet access to internet skills: Digital inequality among older adults. *Universal Access in the Information Society*, *18*(4), 881–890. https://doi.org/10.1007/s10209-018-0617-5

Haritou, M., Anastasiou, A., Kouris, I., Villalonga, S. G., Gancedo, I. O., & Koutsouris, D. (2013). Go-myLife: A context-aware social networking platform adapted to the needs of elderly users. In *Proceedings of the 6th international conference on PErvasive technologies related to assistive environments* (pp. 1–5). https://doi.org/10.1145/2504335.2504343

Harley, D., & Fitzpatrick, G. (2009). Creating a conversational context through video blogging: A case study of Geriatric1927. *Computers in Human Behavior*, *25*(3), 679–689. https://doi.org/10.1016/j.chb.2008.08.011

Harley, D., & Fitzpatrick, G. (n.d.). *Appropriation of social networking by older people: Two case studies.*

Harris, D. A., Thomas, K. S., Epstein-Lubow, G., & Jutkowitz, E. (2020). The association between personal technology use and cognition: Does use matter? *Gerontechnology*, *19*(3), 1–9. https://doi.org/10.4017/gt.2020.19.003.01

Hart, T. A., Chaparro, B. S., & Halcomb, C. G. (2008). Evaluating websites for older adults: Adherence to 'senior-friendly' guidelines and end-user performance. *Behaviour and Information Technology*, *27*(3), 191–199. https://doi.org/10.1080/01449290600802031

Hauser, S., Oogjes, D., Wakkary, R., & Verbeek, P.-P. (2018). An annotated portfolio on doing postphenomenology through research products. In *Proceedings of the 2018 designing interactive systems conference* (pp. 459–471). https://doi.org/10.1145/3196709.3196745

Hays, T., & Minichiello, V. (2005). The meaning of music in the lives of older people: A qualitative study. *Psychology of Music*, *33*(4), 437–451. https://doi.org/10.1177/0305735605056160

Hays, T., Bright, R., & Minichiello, V. (2002). The contribution of music to positive aging: A review. *Journal of Aging and Identity.*

Heart, T., & Kalderon, E. (2013). Older adults: Are they ready to adopt health-related ICT? *International Journal of Medical Informatics*, *82*(11), e209–e231. https://doi.org/10.1016/j.ijmedinf.2011.03.002

Herczeg, M., & Koch, M. (2024). The future of HCI: Editorial. *I-Com*, *23*(2), 131–132. https://doi.org/10.1515/icom-2024-0032

Hermann, J., Oktay, S., Lisetschko, A., & Dogangün, A. (2023). A systematic literature review on the use of social robots in elderly care. In *Proceedings of the 35th Australian computer–human interaction conference* (pp. 221–230). https://doi.org/10.1145/3638380.3638412

Hernández, N., Refugio, C., Tentori, M., Favela, J., & Ochoa, S. (2013). Mobile and context-aware grocery shopping to promote active aging. In C. Nugent, A. Coronato, & J. Bravo (Eds.), *Ambient assisted living and active aging* (Vol. 8277, pp. 71–79). Springer. https://doi.org/10.1007/978-3-319-03092-0_11

Heyn, L., Ellington, L., & Eide, H. (2017). An exploration of how positive emotions are expressed by older people and nurse assistants in homecare visits. *Patient Education and Counseling, 100*(11), 2125–2127. https://doi.org/10.1016/j.pec.2017.05.020

Hill, R., Betts, L. R., & Gardner, S. E. (2015). Older adults' experiences and perceptions of digital technology: (Dis)empowerment, wellbeing, and inclusion. *Computers in Human Behavior, 48*, 415–423. https://doi.org/10.1016/j.chb.2015.01.062

Hjorth, L., Sheahan, J., Figueiredo, B., Martin, D., Reid, M., Aleti, T., & Buschgens, M. (2025). Playing with persona: Highlighting older adults' lived experience with the digital media. *Convergence: The International Journal of Research into New Media Technologies, 31*(1), 368–384. https://doi.org/10.1177/13548565241247415

Hodge, J., Balaam, M., Hastings, S., & Morrissey, K. (2018). Exploring the design of tailored virtual reality experiences for people with dementia. In *Proceedings of the 2018 CHI conference on human factors in computing systems* (pp. 1–13). https://doi.org/10.1145/3173574.3174088

Hofer, S. M., & Alwin, D. F. (Eds.). (2008). *Handbook of cognitive aging: Interdisciplinary perspectives*. Sage Publications.

Hollinworth, N., & Hwang, F. (2009). Learning how older adults undertake computer tasks. In *Proceedings of the 11th international ACM SIGACCESS conference on computers and accessibility* (pp. 245–246). https://doi.org/10.1145/1639642.1639697

Holttum, S. (2016). Do computers increase older people's inclusion and wellbeing? *Mental Health and Social Inclusion, 20*(1), 6–11. https://doi.org/10.1108/MHSI-11-2015-0041

Holzinger, A., & Miesenberger, K. (2009). *HCI and usability for e-inclusion: 5th symposium of the workgroup human–computer interaction and usability engineering of the Austrian computer society, USAB 2009, Linz, Austria, November 9–10, 2009 proceedings*. Springer. https://doi.org/10.1007/978-3-642-10308-7

Höök, K., Dalsgaard, P., Reeves, S., Bardzell, J., Löwgren, J., Stolterman, E., & Rogers, Y. (2015). Knowledge production in interaction design. In *Proceedings of the 33rd annual ACM conference extended abstracts on human factors in computing systems* (pp. 2429–2432). https://doi.org/10.1145/2702613.2702653

Horstmann, A. C., Schubert, T., Lambrich, L., & Strathmann, C. (2023). Alexa, i do not want to be patronized: A qualitative interview study to explore older adults' attitudes towards intelligent voice assistants. In *Proceedings of the 23rd ACM international conference on intelligent virtual agents* (pp. 1–10). https://doi.org/10.1145/3570945.3607342

Hourcade, J. P., Nguyen, C. M., Perry, K. B., & Denburg, N. L. (2010). Pointassist for older adults: Analyzing sub-movement characteristics to aid in pointing tasks.

In *Proceedings of the SIGCHI conference on human factors in computing systems* (pp. 1115–1124). https://doi.org/10.1145/1753326.1753494

Hsu, E. L., Elliott, A., Ishii, Y., Sawai, A., & Katagiri, M. (2020). The development of aged care robots in Japan as a varied process. *Technology in Society, 63*, 101366. https://doi.org/10.1016/j.techsoc.2020.101366

Hsu, L.-J., & Chung, C.-F. (2024). Dancing with the roles: Towards designing technology that supports the multifaceted roles of caregivers for older adults. In *Proceedings of the CHI conference on human factors in computing systems* (pp. 1–12). https://doi.org/10.1145/3613904.3642728

Hu, Y., Qu, Y., Maus, A., & Mutlu, B. (2022). Polite or direct? Conversation design of a smart display for older adults based on politeness theory. In *CHI conference on human factors in computing systems* (pp. 1–15). https://doi.org/10.1145/3491102.3517525

Huang, J., Tan, Q., & Fang, Q. (2024). VitalStep: Revitalizing elderly health and connectivity with e-square-dance. In *Extended abstracts of the CHI conference on human factors in computing systems* (pp. 1–7). https://doi.org/10.1145/3613905.3647978

Hunsaker, A., Nguyen, M. H., Fuchs, J., Djukaric, T., Hugentobler, L., & Hargittai, E. (2019). "He explained it to me and i also did it myself": How older adults get support with their technology uses. *Socius: Sociological Research for a Dynamic World, 5*, 2378023119887866. https://doi.org/10.1177/2378023119887866

Hwang, J. (n.d.). *Exploring lived experiences of instrumental learning and playing after retirement.*

Ineichen, C., Biller-Andorno, N., & Deplazes-Zemp, A. (2021). Between fascination and concern: An exploratory study of senior citizens' attitudes towards synthetic biology and agricultural biotechnology. *Universal Access in the Information Society, 20*(2), 391–404. https://doi.org/10.1007/s10209-020-00719-6

Iñiguez-Berrozpe, T., Valero-Errazu, D., & Elboj-Saso, C. (2018). Hacia una Sociedad de la Información inclusiva. Competencia tecnológica y habilidades relacionadas con las Tecnologías de la Información y la Comunicación (TIC) de los adultos maduros. *Revista Mediterránea de Comunicación, 9*(2), 25. https://doi.org/10.14198/MEDCOM2018.9.2.9

Interactions—January/February 2024. (n.d.).

Internet users in the UK. (n.d.).

Ioannidis, J. P. A., Fanelli, D., Dunne, D. D., & Goodman, S. N. (2015). Meta-research: Evaluation and improvement of research methods and practices. *PLoS Biology, 13*(10), e1002264. https://doi.org/10.1371/journal.pbio.1002264

Irving, P. (n.d.). *The upside of aging.*

Ito, M., O'Day, V. L., Adler, A., Linde, C., & Mynatt, E. D. (n.d.). *Making a place for seniors on the net: SeniorNet, senior identity, and the digital divide.*

Jakob, D., Wilhelm, S., Gerl, A., & Ahrens, D. (2021). A quantitative study on awareness, usage and reservations of voice control interfaces by elderly people. In

C. Stephanidis, D. Harris, W.-C. Li, D. D. Schmorrow, C. M. Fidopiastis, M. Antona, Q. Gao, J. Zhou, P. Zaphiris, A. Ioannou, R. A. Sottilare, J. Schwarz, & M. Rauterberg (Eds.), *HCI international 2021—late breaking papers: Cognition, inclusion, learning, and culture* (Vol. 13096, pp. 237–257). Springer. https://doi.org/10.1007/978-3-030-90328-2_15

Jakob, D., Wilhelm, S., Gerl, A., Ahrens, D., & Wahl, F. (2025). Adapting voice assistant technology for older adults: A comprehensive study on usability, learning patterns, and acceptance. *Digital, 5*(1), 4. https://doi.org/10.3390/digital5010004

Jarke, J. (2019). Open government for all? Co-creating digital public services for older adults through data walks. *Online Information Review, 43*(6), 1003–1020. https://doi.org/10.1108/OIR-02-2018-0059

Jarke, J. (2021). *Co-creating digital public services for an ageing society: Evidence for user-centric design* (Vol. 6). Springer. https://doi.org/10.1007/978-3-030-52873-7

Jarke, J., Gerhard, U., & Kubicek, H. (2019). *Text/conference paper*. https://doi.org/10.18420/INF2019_08

Jastrzembski, T. S., & Charness, N. (2007). The model human processor and the older adult: Parameter estimation and validation within a mobile phone task. *Journal of Experimental Psychology: Applied, 13*(4), 224–248. https://doi.org/10.1037/1076-898X.13.4.224

Jegundo, A. L., Dantas, C., Quintas, J., Dutra, J., Almeida, A. L., Caravau, H., Rosa, A. F., Martins, A. I., & Pacheco Rocha, N. (2020). Perceived usefulness, satisfaction, ease of use and potential of a virtual companion to support the care provision for older adults. *Technologies, 8*(3), 42. https://doi.org/10.3390/technologies8030042

Jelen, B., Monsey, S., & Siek, K. A. (2019). Older adults as makers of custom electronics: Iterating on craftec. In *Extended abstracts of the 2019 CHI conference on human factors in computing systems* (pp. 1–6). https://doi.org/10.1145/3290607.3312755

Jin, T., Wang, Y., & Zhang, W. (2023). Exploring older people's preferences in the design features of virtual agents: A comprehensive analysis. In *Proceedings of the eleventh international symposium of Chinese CHI* (pp. 182–196). https://doi.org/10.1145/3629606.3629624

Jin, X. (2024). Empowering autonomous digital learning for older adults. *Extended abstracts of the CHI conference on human factors in computing systems* (pp. 1–6). https://doi.org/10.1145/3613905.3651133

Jin, X., Kuang, E., & Fan, M. (2021). "Too old to bank digitally?": A survey of banking practices and challenges among older adults in China. *Designing Interactive Systems Conference 2021*, 802–814. https://doi.org/10.1145/3461778.3462127

Jin, Y., Cai, W., Chen, L., Zhang, Y., Doherty, G., & Jiang, T. (2024). Exploring the design of generative AI in supporting music-based reminiscence for older adults. In *Proceedings of the CHI conference on human factors in computing systems* (pp. 1–17). https://doi.org/10.1145/3613904.3642800

Johfre, S. (2020). What age is in a name? *Sociological Science, 7*, 367–390. https://doi.org/10.15195/v7.a15

Jolanki, O. H. (2009). Agency in talk about old age and health. *Journal of Aging Studies, 23*(4), 215–226. https://doi.org/10.1016/j.jaging.2007.12.020

Jones, I. R. (Ed.). (2008). *Ageing in a consumer society: From passive to active consumption in Britain.* Policy Press.

Joyce, K. A., & Loe, M. (2010). *Technogenarians.*

Jung, E. H., & Sundar, S. S. (2016). Senior citizens on Facebook: How do they interact and why? *Computers in Human Behavior, 61*, 27–35. https://doi.org/10.1016/j.chb.2016.02.080

Jutai, J. W., Hatoum, F., Bhardwaj, D., & Hosseini, M. (2024). Implementation of digital health technologies for older adults: A scoping review. *Frontiers in Aging, 5*, 1349520. https://doi.org/10.3389/fragi.2024.1349520

Kandappu, T., Subbaraju, V., & Xu, Q. (2021). PrivacyPrimer: Towards privacy-preserving episodic memory support for older adults. *Proceedings of the ACM on Human–Computer Interaction, 5*, 1–32. https://doi.org/10.1145/3476047

Kang, N. E., & Yoon, W. C. (2008). Age- and experience-related user behavior differences in the use of complicated electronic devices. *International Journal of Human–Computer Studies, 66*(6), 425–437. https://doi.org/10.1016/j.ijhcs.2007.12.003

Kania-Lundholm, M., & Torres, S. (2015). The divide within: Older active ICT users position themselves against different 'Others'. *Journal of Aging Studies, 35*, 26–36. https://doi.org/10.1016/j.jaging.2015.07.008

Kantner, L., & Rosenbaum, S. (2003). Usable computers for the elderly: Applying coaching experiences. In *IEEE international professional communication conference, 2003. IPCC 2003. Proceedings* (p. 10). https://doi.org/10.1109/IPCC.2003.1245476

Karaosmanoglu, S., Kruse, L., Rings, S., & Steinicke, F. (2022). Canoe VR: An immersive exergame to support cognitive and physical exercises of older adults. In *CHI conference on human factors in computing systems extended abstracts* (pp. 1–7). https://doi.org/10.1145/3491101.3519736

Katz, S., & Marshall, B. L. (2018). Tracked and fit: FitBits, brain games, and the quantified aging body. *Journal of Aging Studies, 45*, 63–68. https://doi.org/10.1016/j.jaging.2018.01.009

Khamidullina, Z. (2018). *Community-building in later life and the role of information and communication technologies.*

Khanuja, N. M. (2025). Designing technological interventions to enhance self-perceptions of aging (SPA) and well-being of older adults in India. In *Companion publication of the 2025 ACM designing interactive systems conference* (pp. 111–115). https://doi.org/10.1145/3715668.3735620

Khoo, W., Hsu, L.-J., Amon, K. J., Chakilam, P. V., Chen, W.-C., Kaufman, Z., Lungu, A., Sato, H., Seliger, E., Swaminathan, M., Tsui, K. M., Crandall, D. J., & Šabanović, S. (2023). Spill the tea: When robot conversation agents support well-being for older adults. In *Companion of the 2023 ACM/IEEE international conference on human–robot interaction* (pp. 178–182). https://doi.org/10.1145/3568294.3580067

Khvorostianov, N., Elias, N., & Nimrod, G. (2012). 'Without it I am nothing': The internet in the lives of older immigrants. *New Media and Society*, *14*(4), 583–599. https://doi.org/10.1177/1461444811421599

Kilpeläinen, A., & Seppänen, M. (2014). Information technology and everyday life in ageing rural villages. *Journal of Rural Studies*, *33*, 1–8. https://doi.org/10.1016/j.jrustud.2013.10.005

Kim, S., Shin, D., Kim, J., Kwon, S., & Kim, J. (2023). How older adults use online videos for learning. In *Proceedings of the 2023 CHI conference on human factors in computing systems* (pp. 1–16). https://doi.org/10.1145/3544548.3580671

Kivimäki, T., Liolis, K., Yildizoglu, U., Kaila, L., Vainio, A.-M., Konakas, S., Katsiouli, P., Pensas, H., Summanen, K., Pantazis, S., Moisio, H., Andrikopoulos, I., & Vanhala, J. (2012). On an advanced ICT-enabled system for the social inclusion of the elderly. In *Proceedings of the 5th international conference on PErvasive technologies related to assistive environments* (pp. 1–6). https://doi.org/10.1145/2413097.2413115

Klimova, B., Simonova, I., Poulova, P., Truhlarova, Z., & Kuca, K. (n.d.). *Older people and their attitude to the use of information and communication technologies: A review study with special focus on the Czech Republic (Older people and their attitude to ICT)*.

Knowles, B., Singh, A., Ambe, A. H., Brewer, R. N., Lazar, A., Petrie, H., Vines, J., & Waycott, J. (2024). HCI and aging: New directions, new principles. In *Extended abstracts of the CHI conference on human factors in computing systems* (pp. 1–5). https://doi.org/10.1145/3613905.3636295

Kobayashi, M., Hiyama, A., Miura, T., Asakawa, C., Hirose, M., & Ifukube, T. (2011). Elderly user evaluation of mobile touchscreen interactions. In P. Campos, N. Graham, J. Jorge, N. Nunes, P. Palanque, & M. Winckler (Eds.), *Human–computer interaction: INTERACT 2011* (Vol. 6946, pp. 83–99). Springer. https://doi.org/10.1007/978-3-642-23774-4_9

Komp, K., & Aartsen, M. (Eds.). (2013). *Old age in Europe: A textbook of gerontology*. Springer. https://doi.org/10.1007/978-94-007-6134-6

Kooij, D., De Lange, A., Jansen, P., & Dikkers, J. (2008). Older workers' motivation to continue to work: Five meanings of age: A conceptual review. *Journal of Managerial Psychology*, *23*(4), 364–394. https://doi.org/10.1108/02683940810869015

Kopeć, W., Wichrowski, M., Kalinowski, K., Jaskulska, A., Skorupska, K., Cnotkowski, D., Tyszka, J., Popieluch, A., Voitenkova, A., Masłyk, R., Gago, P., Krzywicki, M., Kornacka, M., Biele, C., Kobyliński, P., Kowalski, J., Abramczuk, K., Zdrodowska, A., Pochwatko, G., et al. (2019). *VR with older adults: Participatory design of a virtual ATM training simulation* (No. arXiv:1911.00466). arXiv. https://doi.org/10.48550/arXiv.1911.00466

Kuang, E. (2025). Evaluating usability challenges in VR games for older adults: A comparison with and without AI assistance. In *Adjunct proceedings of the 38th annual*

ACM symposium on user interface software and technology (pp. 1–3). https://doi.org/10.1145/3746058.3758343

Kubik, S. (2009). Motivations for cell phone use by older Americans. *Gerontechnology, 8*(3), 150–164. https://doi.org/10.4017/gt.2009.08.03.007.00

Kuijer, L., & Giaccardi, E. (2018). Co-performance: Conceptualizing the role of artificial agency in the design of everyday life. In *Proceedings of the 2018 CHI conference on human factors in computing systems* (pp. 1–13). https://doi.org/10.1145/3173574.3173699

Kuoppamäki, S., Jaberibraheem, R., Hellstrand, M., & McMillan, D. (2023). Designing multi-modal conversational agents for the kitchen with older adults: A participatory design study. *International Journal of Social Robotics, 15*(9–10), 1507–1523. https://doi.org/10.1007/s12369-023-01055-4

Lafontaine, C., Sawchuk, K., & DeJong, S. (2020). Social justice games: Building an escape room on elder abuse through participatory action research. *The Computer Games Journal, 9*(2), 189–205. https://doi.org/10.1007/s40869-020-00105-5

Langston, E., Charness, N., & Boot, W. (2024). Are virtual assistants trustworthy for medicare information: An examination of accuracy and reliability. *The Gerontologist, 64*(8), gnae062. https://doi.org/10.1093/geront/gnae062

Lappa, D. (2018). *Facebook and Greek retirees' relationships.*

Laslett, P. (2000). Review reviewed work(s): Age power: How the twenty-first century will be ruled by the new old by Ken Dychtwald. *Population and Development Review, 26*(2), 398–399.

Laukka, P. (2007). Uses of music and psychological well-being among the elderly. *Journal of Happiness Studies, 8*(2), 215. https://doi.org/10.1007/s10902-006-9024-3

Lazar, A., Jelen, B., Pradhan, A., & Siek, K. A. (2021). Adopting diffractive reading to advance HCI research: A case study on technology for aging. *ACM Transactions on Computer–Human Interaction, 28*(5), 1–29. https://doi.org/10.1145/3462326

Lazar, A., Pradhan, A., Jelen, B., A. Siek, K., & Leitch, A. (2021). Studying the formation of an older adult-led makerspace. In *Proceedings of the 2021 CHI conference on human factors in computing systems* (pp. 1–11). https://doi.org/10.1145/3411764.3445146

Lazar, A., Thompson, H. J., Piper, A. M., & Demiris, G. (2016). Rethinking the design of robotic pets for older adults. In *Proceedings of the 2016 ACM conference on designing interactive systems* (pp. 1034–1046). https://doi.org/10.1145/2901790.2901811

Leamy, M., & Clough, R. (2006). *How older people became researchers: Training, guidance and practice in action* (Electronic version). Joseph Rowntree Foundation.

Lee, C. C., Czaja, S. J., Moxley, J. H., Sharit, J., Boot, W. R., Charness, N., & Rogers, W. A. (2019). Attitudes toward computers across adulthood from 1994 to 2013. *The Gerontologist, 59*(1), 22–33. https://doi.org/10.1093/geront/gny081

Lee, H. R., & Riek, L. (2023). Designing robots for aging: Wisdom as a critical lens. *ACM Transactions on Human–Robot Interaction, 12*(1), 1–21. https://doi.org/10.1145/3549531

Leonardi, C., Albertini, A., Pianesi, F., & Zancanaro, M. (2010). An exploratory study of a touch-based gestural interface for elderly. In *Proceedings of the 6th Nordic conference on human–computer interaction: Extending boundaries* (pp. 845–850). https://doi.org/10.1145/1868914.1869045

Lepicard, G., & Vigouroux, N. (2012). Comparison between single-touch and multi-touch interaction for older people. In K. Miesenberger, A. Karshmer, P. Penaz, & W. Zagler (Eds.), *Computers helping people with special needs* (Vol. 7382, pp. 658–665). Springer. https://doi.org/10.1007/978-3-642-31522-0_99

Leung, R., Tang, C., Haddad, S., Mcgrenere, J., Graf, P., & Ingriany, V. (2012). How older adults learn to use mobile devices: Survey and field investigations. *ACM Transactions on Accessible Computing, 4*(3), 1–33. https://doi.org/10.1145/2399193.2399195

Li, G., & Seaborn, K. (2024). No joke: An embodied conversational agent greeting older adults with humour or a smile unrelated to initial acceptance. In *Extended abstracts of the CHI conference on human factors in computing systems* (pp. 1–7). https://doi.org/10.1145/3613905.3650918

Li, G., & Tang, T. (2025). Online performance and interface design implications among older adults: A systematic review of eye tracking studies. *Applied Ergonomics, 128*, 104538. https://doi.org/10.1016/j.apergo.2025.104538

Li, J., Zong, B., Cheng, T., Li, Y., Mynatt, E. D., & Dhekne, A. (2023). Privacy versus awareness: Relieving the tension between older adults and adult children when sharing in-home activity data. *Proceedings of the ACM on Human–Computer Interaction, 7*, 1–30. https://doi.org/10.1145/3610202

Li, X., Ren, X., Suzuki, X., Yamaji, N., Fung, K. W., & Gondo, Y. (2024). Designing a multisensory VR game prototype for older adults: The acceptability and design implications. In *Proceedings of the CHI conference on human factors in computing systems* (pp. 1–18). https://doi.org/10.1145/3613904.3642948

Lian, J.-W., & Yen, D. C. (2014). Online shopping drivers and barriers for older adults: Age and gender differences. *Computers in Human Behavior, 37*, 133–143. https://doi.org/10.1016/j.chb.2014.04.028

Liddle, J. L. M., Parkinson, L., & Sibbritt, D. W. (2013). Purpose and pleasure in late life: Conceptualising older women's participation in art and craft activities. *Journal of Aging Studies, 27*(4), 330–338. https://doi.org/10.1016/j.jaging.2013.08.002

Liddle, J., Pitcher, N., Montague, K., Hanratty, B., Standing, H., & Scharf, T. (2020). Connecting at local level: Exploring opportunities for future design of technology to support social connections in age-friendly communities. *International Journal of Environmental Research and Public Health, 17*(15), 5544. https://doi.org/10.3390/ijerph17155544

Lim, C. S. C. (2010). Designing inclusive ICT products for older users: Taking into account the technology generation effect. *Journal of Engineering Design, 21*(2–3), 189–206. https://doi.org/10.1080/09544820903317001

Liu, Y., & Lachman, M. E. (2021). A group-based walking study to enhance physical activity among older adults: The role of social engagement. *Research on Aging, 43*(9–10), 368–377. https://doi.org/10.1177/0164027520963613

Lloyd, L., Calnan, M., Cameron, A., Seymour, J., & Smith, R. (2014). Identity in the fourth age: Perseverance, adaptation and maintaining dignity. *Ageing and Society, 34*(1), 1–19. https://doi.org/10.1017/S0144686X12000761

Long, K. M., Casey, K., Bhar, S., Al Mahmud, A., Curran, S., Hunter, K., & Lim, M. H. (2024). Understanding perspectives of older adults on the role of technology in the wider context of their social relationships. *Ageing and Society, 44*(6), 1453–1476. https://doi.org/10.1017/S0144686X2200085X

López Gómez, D. (2015). Little arrangements that matter. Rethinking autonomy-enabling innovations for later life. *Technological Forecasting and Social Change, 93*, 91–101. https://doi.org/10.1016/j.techfore.2014.02.015

Loup, J., Subasi, Ö., & Fitzpatrick, G. (2017). Aging, HCI, and personal perceptions of time. In *Proceedings of the 2017 CHI conference extended abstracts on human factors in computing systems* (pp. 1853–1860). https://doi.org/10.1145/3027063.3053079

Lunn, D., & Harper, S. (2011). Providing assistance to older users of dynamic web content. *Computers in Human Behavior, 27*(6), 2098–2107. https://doi.org/10.1016/j.chb.2011.06.004

Ma, Q., Chan, A. H. S., & Teh, P.-L. (2020). Bridging the digital divide for older adults via observational training: Effects of model identity from a generational perspective. *Sustainability, 12*(11), 4555. https://doi.org/10.3390/su12114555

Madden, M. (n.d.). *Older adults and social media.*

Maestre, J. F., Kresnye, K. C., Dunbar, J. C., Connelly, C. L., Siek, K. A., & Shih, P. C. (2020). Conducting HCI research with people living with HIV remotely: Lessons learned and best practices. In *Extended abstracts of the 2020 CHI conference on human factors in computing systems* (pp. 1–8). https://doi.org/10.1145/3334480.3375202

Malik, J., N. Di Napoli Parr, M., Flathau, J., Tang, H., K. Kearney, J., M. Plumert, J., & Rector, K. (2021). Determining the effect of smartphone alerts and warnings on street-crossing behavior in non-mobility-impaired older and younger adults. In *Proceedings of the 2021 CHI conference on human factors in computing systems* (pp. 1–12). https://doi.org/10.1145/3411764.3445234

Malta, S., & Farquharson, K. (n.d.). *Old dogs, new tricks? Online dating and older adults.*

Mannheim, I., & Köttl, H. (2024). Ageism and (successful) digital engagement: A proposed theoretical model. *The Gerontologist, 64*(9), gnae078. https://doi.org/10.1093/geront/gnae078

Mao, M., Blackwell, A. F., & Good, D. A. (2015). Music in the retiring life: A review of evaluation methods and potential factors. In J. Zhou & G. Salvendy (Eds.), *Human aspects of IT for the aged population. Design for aging* (Vol. 9193, pp. 73–83). Springer. https://doi.org/10.1007/978-3-319-20892-3_8

Marquié, J. C., Jourdan-Boddaert, L., & Huet, N. (2002). Do older adults underestimate their actual computer knowledge? *Behaviour and Information Technology, 21*(4), 273–280. https://doi.org/10.1080/0144929021000020998

Marston, H. R., Genoe, R., Freeman, S., Kulczycki, C., & Musselwhite, C. (2019). Older adults' perceptions of ICT: Main findings from the technology in later life (TILL) study. *Healthcare, 7*(3), 86. https://doi.org/10.3390/healthcare7030086

Martin-Hammond, A., Vemireddy, S., & Rao, K. (2018). Engaging older adults in the participatory design of intelligent health search tools. In *Proceedings of the 12th EAI international conference on pervasive computing technologies for healthcare* (pp. 280–284). https://doi.org/10.1145/3240925.3240972

Martínez Maroto, A. (2009). *Nuevas miradas sobre el envejecimiento* (1a. ed.). IMSERSO.

Massimi, M., Odom, W., Banks, R., & Kirk, D. (2011). Matters of life and death: Locating the end of life in lifespan-oriented HCI research. In *Proceedings of the SIGCHI conference on human factors in computing systems* (pp. 987–996). https://doi.org/10.1145/1978942.1979090

Mayer, P., & Panek, P. (2017). Involving older and vulnerable persons in the design process of an enhanced toilet system. In *Proceedings of the 2017 CHI conference extended abstracts on human factors in computing systems* (pp. 2774–2780). https://doi.org/10.1145/3027063.3053178

Mcdonald, N., & Mentis, H. M. (2021). "Citizens too": Safety setting collaboration among older adults with memory concerns. *ACM Transactions on Computer–Human Interaction, 28*(5), 1–32. https://doi.org/10.1145/3465217

McGlynn, S. A., Kemple, S., Mitzner, T. L., King, C.-H. A., & Rogers, W. A. (2017). Understanding the potential of PARO for healthy older adults. *International Journal of Human–Computer Studies, 100*, 33–47. https://doi.org/10.1016/j.ijhcs.2016.12.004

McGrath, C., Molinaro, M. L., Sheldrake, E. J., Laliberte Rudman, D., & Astell, A. (2019). A protocol paper on the preservation of identity: Understanding the technology adoption patterns of older adults with age-related vision loss (ARVL). *International Journal of Qualitative Methods, 18*, 1609406919831833. https://doi.org/10.1177/1609406919831833

McLaughlin, A. C., Rogers, W. A., & Fisk, A. D. (2009). Using direct and indirect input devices: Attention demands and age-related differences. *ACM Transactions on Computer–Human Interaction, 16*(1), 1–15. https://doi.org/10.1145/1502800.1502802

McLean, A. (2011). Ethical frontiers of ICT and older users: Cultural, pragmatic and ethical issues. *Ethics and Information Technology, 13*(4), 313–326. https://doi.org/10.1007/s10676-011-9276-4

McNeill, A. R., Coventry, L., Pywell, J., & Briggs, P. (2017). Privacy considerations when designing social network systems to support successful ageing. In *Proceedings of the 2017 CHI conference on human factors in computing systems* (pp. 6425–6437). https://doi.org/10.1145/3025453.3025861

Means, R., & Evans, S. (2012). Communities of place and communities of interest? An exploration of their changing role in later life. *Ageing and Society, 32*(8), 1300–1318. https://doi.org/10.1017/S0144686X11000961

Mendel, T., & Toch, E. (2022). Meerkat: A social community support application for older adults. In *CHI conference on human factors in computing systems extended abstracts* (pp. 1–4). https://doi.org/10.1145/3491101.3519909

Menec, V. H., Means, R., Keating, N., Parkhurst, G., & Eales, J. (2011). Conceptualizing age-friendly communities. *Canadian Journal on Aging/La Revue Canadienne Du Vieillissement, 30*(3), 479–493. https://doi.org/10.1017/S0714980811000237

Minichiello, V., Browne, J., & Kendig, H. (2000). Perceptions and consequences of ageism: Views of older people. *Ageing and Society, 20*(3), 253–278. https://doi.org/10.1017/S0144686X99007710

Mitchell, V., Mackley, K. L., Pink, S., Escobar-Tello, C., Wilson, G. T., & Bhamra, T. (2015). Situating digital interventions: Mixed methods for HCI research in the home. *Interacting with Computers, 27*(1), 3–12. https://doi.org/10.1093/iwc/iwu034

Moal-Ulvoas, G. (2017). Positive emotions and spirituality in older travelers. *Annals of Tourism Research, 66*, 151–158. https://doi.org/10.1016/j.annals.2017.07.020

Mody, L., Miller, D. K., McGloin, J. M., Freeman, M., Marcantonio, E. R., Magaziner, J., & Studenski, S. (2008). Recruitment and retention of older adults in aging research: (See editorial comments by Dr. Stephanie Studenski, pp. 2351–2352). *Journal of the American Geriatrics Society, 56*(12), 2340–2348. https://doi.org/10.1111/j.1532-5415.2008.02015.x

Mordini, E., Wright, D., Wadhwa, K., De Hert, P., Mantovani, E., Thestrup, J., Van Steendam, G., D'Amico, A., & Vater, I. (2009). Senior citizens and the ethics of e-inclusion. *Ethics and Information Technology, 11*(3), 203–220. https://doi.org/10.1007/s10676-009-9189-7

Morris, M., Lundell, J., Dishman, E., & Needham, B. (2003). New perspectives on ubiquitous computing from ethnographic study of elders with cognitive decline. In A. K. Dey, A. Schmidt, & J. F. McCarthy (Eds.), *UbiComp 2003: Ubiquitous computing* (Vol. 2864, pp. 227–242). Springer. https://doi.org/10.1007/978-3-540-39653-6_18

Mostaghel, R. (2016). Innovation and technology for the elderly: Systematic literature review. *Journal of Business Research, 69*(11), 4896–4900. https://doi.org/10.1016/j.jbusres.2016.04.049

Mostajeran, F., Steinicke, F., Ariza Nunez, O. J., Gatsios, D., & Fotiadis, D. (2020). Augmented reality for older adults: Exploring acceptability of virtual coaches for home-based balance training in an aging population. In *Proceedings of the 2020 CHI conference on human factors in computing systems* (pp. 1–12). https://doi.org/10.1145/3313831.3376565

Motti, L. G., Vigouroux, N., & Gorce, P. (2013). *Interaction techniques for older adults using touchscreen devices: A literature review.*

Msweli, N. T., & Mawela, T. (2020). Enablers and barriers for mobile commerce and banking services among the elderly in developing countries: A systematic review. In M. Hattingh, M. Matthee, H. Smuts, I. Pappas, Y. K. Dwivedi, & M. Mäntymäki (Eds.), *Responsible design, implementation and use of information and communication technology* (Vol. 12067, pp. 319–330). Springer. https://doi.org/10.1007/978-3-030-45002-1_27

Muller, M., Chilton, L. B., Kantosalo, A., Martin, C. P., & Walsh, G. (2022). GenAICHI: Generative AI and HCI. In *CHI conference on human factors in computing systems extended abstracts* (pp. 1–7). https://doi.org/10.1145/3491101.3503719

Muller, M., Kantosalo, A., Maher, M. L., Martin, C. P., & Walsh, G. (2024). GenAICHI 2024: Generative AI and HCI at CHI 2024. In *Extended abstracts of the CHI conference on human factors in computing systems* (pp. 1–7). https://doi.org/10.1145/3613905.3636294

Murray-Rust, D., Gorkovenko, K., Burnett, D., & Richards, D. (2019). Entangled ethnography: Towards a collective future understanding. In *Proceedings of the halfway to the future symposium 2019* (pp. 1–10). https://doi.org/10.1145/3363384.3363405

Muskens, L., Van Lent, R., Vijfvinkel, A., Van Cann, P., & Shahid, S. (2014). Never too old to use a tablet: Designing tablet applications for the cognitively and physically impaired elderly. In K. Miesenberger, D. Fels, D. Archambault, P. Peňáz, & W. Zagler (Eds.), *Computers helping people with special needs* (Vol. 8547, pp. 391–398). Springer. https://doi.org/10.1007/978-3-319-08596-8_60

Nallam, P., Bhandari, S., Sanders, J., & Martin-Hammond, A. (2020). A question of access: Exploring the perceived benefits and barriers of intelligent voice assistants for improving access to consumer health resources among low-income older adults. *Gerontology and Geriatric Medicine, 6*, 233372142098597. https://doi.org/10.1177/2333721420985975

Namvarpour, M. M., & Razi, A. (2025). The art of talking machines: A comprehensive literature review of conversational user interfaces. In *Proceedings of the 7th ACM conference on conversational user interfaces* (pp. 1–18). https://doi.org/10.1145/3719160.3736621

Nassir, S., & Leong, T. W. (2017). Traversing boundaries: Understanding the experiences of ageing Saudis. In *Proceedings of the 2017 CHI conference on human factors in computing systems* (pp. 6386–6397). https://doi.org/10.1145/3025453.3025618

Naumann, A. B., Wechsung, I., & Hurtienne, J. (2010). Multimodal interaction: A suitable strategy for including older users? *Interacting with Computers, 22*(6), 465–474. https://doi.org/10.1016/j.intcom.2010.08.005

Neef, C., & Richert, A. (2024). Aging with agency: Real-world insights into robot-assisted self-health monitoring for older adults. In *Companion of the 2024 ACM/IEEE international conference on human–robot interaction* (pp. 784–788). https://doi.org/10.1145/3610978.3640561

Neef, C., & Richert, A. (2025). *Likable or intelligent? Comparing social robots and virtual agents for long-term health monitoring.*

Neef, C., Linden, K., & Richert, A. (2024). To be or not to (physically) be? A study on preferences in embodied socially interactive agents for health monitoring of older adults. In *Proceedings of the 2024 33rd IEEE international conference on robot and human interactive communication (ROMAN)* (pp. 1497–1502). https://doi.org/10.1109/RO-MAN60168.2024.10731187

Nehmadi, L., Meyer, J., Parmet, Y., & Ben-Asher, N. (2011). Predicting a screen area's perceived importance from spatial and physical attributes. *Journal of the American Society for Information Science and Technology, 62*(9), 1829–1838. https://doi.org/10.1002/asi.21580

Nelson, Z., & Rui, P. (Eds.). (2008).

Ng, R., & Chow, T. Y. J. (2021). Aging narratives over 210 years (1810–2019). *The Journals of Gerontology: Series B, 76*(9), 1799–1807. https://doi.org/10.1093/geronb/gbaa222

Nicholson, J., Coventry, L., & Briggs, P. (2019). 'If it's important it will be a headline': Cybersecurity information seeking in older adults. In *Proceedings of the 2019 CHI conference on human factors in computing systems* (pp. 1–11). https://doi.org/10.1145/3290605.3300579

Nicolau, H., & Jorge, J. (2012). Elderly text-entry performance on touchscreens. In *Proceedings of the 14th international ACM SIGACCESS conference on computers and accessibility* (pp. 127–134). https://doi.org/10.1145/2384916.2384939

Nimrod, G. (2010). Seniors' online communities: A quantitative content analysis. *The Gerontologist, 50*(3), 382–392. https://doi.org/10.1093/geront/gnp141

Nimrod, G. (2011). The fun culture in seniors' online communities. *The Gerontologist, 51*(2), 226–237. https://doi.org/10.1093/geront/gnq084

Nimrod, G. (2018). Technostress: Measuring a new threat to well-being in later life. *Aging and Mental Health, 22*(8), 1086–1093. https://doi.org/10.1080/13607863.2017.1334037

Nimrod, G. (2020). Aging well in the digital age: Technology in processes of selective optimization with compensation. *The Journals of Gerontology: Series B, 75*(9), 2008–2017. https://doi.org/10.1093/geronb/gbz111

Nimrod, G., & Ivan, L. (2022). The dual roles technology plays in leisure: Insights from a study of grandmothers. *Leisure Sciences, 44*(6), 715–732. https://doi.org/10.1080/01490400.2019.1656123

Nind, M. (2008). *Conducting qualitative research with people with learning, communication and other disabilities: Methodological challenges.*

Norman, G., Bennett, P., & Vardy, E. R. L. C. (2023). Virtual wards: A rapid evidence synthesis and implications for the care of older people. *Age and Ageing, 52*(1), afac319. https://doi.org/10.1093/ageing/afac319

Norval, C., Arnott, J. L., & Hanson, V. L. (2014). What's on your mind? Investigating recommendations for inclusive social networking and older adults. In *Proceedings of the SIGCHI conference on human factors in computing systems* (pp. 3923–3932). https://doi.org/10.1145/2556288.2556992

Nurain, N., Caldeira, C., & Connelly, K. (2021). Older adults' experiences of autonomy during COVID-19 pandemic. In *Extended abstracts of the 2021 CHI conference on human factors in computing systems* (pp. 1–6). https://doi.org/10.1145/3411763.3451674

Nussbaum, J. F., & Coupland, J. (Eds.). (2004). *Handbook of communication and aging research* (2nd ed.). Lawrence Erlbaum Associates.

Nussbaum, J. F., & Coupland, J. (Eds.). (2008). *Handbook of communication and aging research* (2. ed., reprinted). Routledge.

O'Brien, K., Liggett, A., Ramirez-Zohfeld, V., Sunkara, P., & Lindquist, L. A. (2020). Voice-controlled intelligent personal assistants to support aging in place. *Journal of the American Geriatrics Society, 68*(1), 176–179. https://doi.org/10.1111/jgs.16217

O'brien, M. A., Rogers, W. A., & Fisk, A. D. (2012). Understanding age and technology experience differences in use of prior knowledge for everyday technology interactions. *ACM Transactions on Accessible Computing, 4*(2), 1–27. https://doi.org/10.1145/2141943.2141947

O'Connell, C., Quinn, K., Marquez, D., Chin, J., Muramatsu, N., Leiser, S., Gradishar, J., & Desai, S. (2021). Accommodating communication with conversational agents: Examining the perceptions and behaviors of older adults when using voice assistant technology. *AoIR Selected Papers of Internet Research.* https://doi.org/10.5210/spir.v2021i0.12221

Öberg, B.-M. (2017). *Changing worlds and the ageing subject: Dimensions in the study of ageing and later life* (1st ed.). Routledge.

Okonji, P., Lhussier, M., Bailey, C., & Cattan, M. (2015). *Internet use: Perceptions and experiences of visually impaired older adults.*

Oliveira, J., Silva, T., Oliveira, R., & Furtado, E. (2025). The role of embodied conversational agents in supporting older adults with hypertension: A focus group study. In *Proceedings of the 11th international conference on information and communication technologies for ageing well and e-health* (pp. 433–442). https://doi.org/10.5220/0013504300003938

Olsson, T., & Väänänen, K. (2021). How does AI challenge design practice? *Interactions, 28*(4), 62–64. https://doi.org/10.1145/3467479

Olwal, A., Lachanas, D., & Zacharouli, E. (2011). OldGen: Mobile phone personalization for older adults. In *Proceedings of the SIGCHI conference on human factors in computing systems* (pp. 3393–3396). https://doi.org/10.1145/1978942.1979447

Omotayo, F. O., & Akinyode, T. A. (2020). Digital inclusion and the elderly: The case of internet banking use and non-use among older adults in Ekiti State, Nigeria. *Covenant Journal of Business and Social Sciences, 11*(1). https://doi.org/10.47231/EDJ U4275

Oomes, A. H. J., Bojic, M., & Bazen, G. (2009). Supporting cognitive collage creation for pedestrian navigation. In D. Harris (Ed.), *Engineering psychology and cognitive ergonomics* (Vol. 5639, pp. 111–119). Springer. https://doi.org/10.1007/978-3-642-02728-4_12

Osborne, J. W. (n.d.). *Psychological effects of the transition to retirement effects psychologiques de la transition vers la retraite.*

Östlund, B. (2008). The revival of research circles: Meeting the needs of modern aging and the third age. *Educational Gerontology, 34*(4), 255–266. https://doi.org/10.1080/03601270701835916

Otjacques, B., Krier, M., Feltz, F., Ferring, D., & Hoffmann, M. (2010). Designing for older people: A case study in a retirement home. In G. Leitner, M. Hitz, & A. Holzinger (Eds.), *HCI in work and learning, life and leisure* (Vol. 6389, pp. 177–194). Springer. https://doi.org/10.1007/978-3-642-16607-5_11

Pachis, J. A., & Zonneveld, K. L. M. (2019). Comparison of prompting procedures to teach internet skills to older adults. *Journal of Applied Behavior Analysis, 52*(1), 173–187. https://doi.org/10.1002/jaba.519

Padrón Nápoles, V. M., Gachet Páez, D., Esteban Penelas, J. L., García Pérez, O., García Santacruz, M. J., & Martín De Pablos, F. (2020). Smart bus stops as interconnected public spaces for increasing social inclusiveness and quality of life of elder users. *Smart Cities, 3*(2), 430–443. https://doi.org/10.3390/smartcities3020023

Pak, R., & Price, M. M. (2008). Designing an information search interface for younger and older adults. *Human Factors: The Journal of the Human Factors and Ergonomics Society, 50*(4), 614–628. https://doi.org/10.1518/001872008X312314

Palivcová, D., Macík, M., & Míkovec, Z. (2020). Interactive tactile map as a tool for building spatial knowledge of visually impaired older adults. In *Extended abstracts of the 2020 CHI conference on human factors in computing systems* (pp. 1–9). https://doi.org/10.1145/3334480.3382912

Papa, F., Sapio, B., & Pelagalli, M. F. (n.d.). *User experience of elderly people with digital television: A qualitative investigation.*

Park, C., Cho, M., Shin, M., Ryu, J.-K., & Jang, M. (2025). *Adaptive robot-mediated assessment using LLM for enhanced survey quality in older adults care programs.*

Parra, C., D'Andrea, V., & Giacomin, G. (n.d.). *Enabling community participation of senior citizens through participatory design and ICT.*

Parsons, S., Yuill, N., Good, J., & Brosnan, M. (2020). 'Whose agenda? Who knows best? Whose voice?' Co-creating a technology research roadmap with autism stakeholders. *Disability and Society, 35*(2), 201–234. https://doi.org/10.1080/09687599.2019.1624152

Patrick Rau, P.-L., & Hsu, J.-W. (2005). Interaction devices and web design for novice older users. *Educational Gerontology, 31*(1), 19–40. https://doi.org/10.1080/036012705 90522170

Patsoule, E., & Koutsabasis, P. (2012). Redesigning web sites for older adults. In *Proceedings of the 5th international conference on PErvasive technologies related to assistive environments* (pp. 1–8). https://doi.org/10.1145/2413097.2413114

Pedell, S., Beh, J., Mozuna, K., & Duong, S. (2013). Engaging older adults in activity group settings playing games on touch tablets. In *Proceedings of the 25th Australian computer–human interaction conference: augmentation, application, innovation, collaboration* (pp. 477–480). https://doi.org/10.1145/2541016.2541090

Pedell, S., Vetere, F., Kulik, L., Ozanne, E., & Gruner, A. (n.d.). *Social isolation of older people: The role of domestic technologies.*

Peek, S. T. M., Luijkx, K. G., Rijnaard, M. D., Nieboer, M. E., Van Der Voort, C. S., Aarts, S., Van Hoof, J., Vrijhoef, H. J. M., & Wouters, E. J. M. (2016). Older adults' reasons for using technology while aging in place. *Gerontology, 62*(2), 226–237. https://doi.org/10.1159/000430949

Peine, A., Rollwagen, I., & Neven, L. (2014). The rise of the "innosumer"—Rethinking older technology users. *Technological Forecasting and Social Change, 82*, 199–214. https://doi.org/10.1016/j.techfore.2013.06.013

Peine, A., Van Cooten, V., & Neven, L. (2017). Rejuvenating design: Bikes, batteries, and older adopters in the diffusion of e-bikes. *Science, Technology, and Human Values, 42*(3), 429–459. https://doi.org/10.1177/0162243916664589

Peral-Peral, B., Villarejo-Ramos, Á. F., & Arenas-Gaitán, J. (2020). Self-efficacy and anxiety as determinants of older adults' use of internet banking services. *Universal Access in the Information Society, 19*(4), 825–840. https://doi.org/10.1007/s10209-019-00691-w

Petrie, H., & Darzentas, J. (2017). Older people and robotic technologies in the home: Perspectives from recent research literature. In *Proceedings of the 10th international conference on PErvasive technologies related to assistive environments* (pp. 29–36). https://doi.org/10.1145/3056540.3056553

Petrie, H., & Weber, G. (2016). Technology for disabled and older people: What have we achieved, where are we going? In *Proceedings of the 2016 CHI conference extended abstracts on human factors in computing systems* (pp. 1085–1087). https://doi.org/10.1145/2851581.2886443

Petrie, H., Gallagher, B., & Darzentas, J. S. (2014). A critical review of eight years of research on technologies for disabled and older people. In K. Miesenberger, D. Fels, D. Archambault, P. Peňáz, & W. Zagler (Eds.), *Computers helping people with special needs* (Vol. 8548, pp. 260–266). Springer. https://doi.org/10.1007/978-3-319-08599-9_40

Petrovcic, A., Boot, W. R., Burnik, T., & Dolnicar, V. (2019). Improving the measurement of older adults' mobile device proficiency: Results and implications from a study of older adult smartphone users. *IEEE Access*, *7*, 150412–150422. https://doi.org/10.1109/ACCESS.2019.2947765

Petrovčič, A., Vehovar, V., & Dolničar, V. (2016). Landline and mobile phone communication in social companionship networks of older adults: An empirical investigation in Slovenia. *Technology in Society*, *45*, 91–102. https://doi.org/10.1016/j.techsoc.2016.02.007

Phillips, J., Ajrouch, K., & Hillcoat-Nallétamby, S. (2010). *Key concepts in social gerontology*. SAGE Publications Ltd. https://doi.org/10.4135/9781446251058

Phillipson, C., Downs, M., Vincent, J. A., & British Society of Gerontology (Eds.). (2006). *The futures of old age*. SAGE Publications.

Pickard, S. (2009). Governing old age: The 'case managed' older person. *Sociology*, *43*(1), 67–84. https://doi.org/10.1177/0038038508099098

Piper, A. M., Brewer, R., & Cornejo, R. (2017). Technology learning and use among older adults with late-life vision impairments. *Universal Access in the Information Society*, *16*(3), 699–711. https://doi.org/10.1007/s10209-016-0500-1

Piper, A. M., Campbell, R., & Hollan, J. D. (2010). Exploring the accessibility and appeal of surface computing for older adult health care support. In *Proceedings of the SIGCHI conference on human factors in computing systems* (pp. 907–916). https://doi.org/10.1145/1753326.1753461

Plaza, I., Martín, L., Martin, S., & Medrano, C. (2011). Mobile applications in an aging society: Status and trends. *Journal of Systems and Software*, *84*(11), 1977–1988. https://doi.org/10.1016/j.jss.2011.05.035

Podgórniak-Krzykacz, A., Przywojska, J., & Wiktorowicz, J. (2020). Smart and age-friendly communities in Poland. An analysis of institutional and individual conditions for a new concept of smart development of ageing communities. *Energies*, *13*(9), 2268. https://doi.org/10.3390/en13092268

Pollmann, K. (2021). The modality card deck: Co-creating multi-modal behavioral expressions for social robots with older adults. *Multimodal Technologies and Interaction*, *5*(7), 33. https://doi.org/10.3390/mti5070033

Pradhan, A., Erete, S., Chopra, S., Upadhyay, P., Sule, O., & Lazar, A. (2025). 'No, not that voice again!': Engaging older adults in design of anthropomorphic voice assistants. In *Proceedings of the ACM on human–computer interaction*, *9*(2), 1–30. https://doi.org/10.1145/3711039

Pradhan, A., Jelen, B., Siek, K. A., Chan, J., & Lazar, A. (2020). Understanding older adults' participation in design workshops. In *Proceedings of the 2020 CHI conference on human factors in computing systems* (pp. 1–15). https://doi.org/10.1145/3313831.3376299

Price, M. M., Pak, R., Müller, H., & Stronge, A. (2013). Older adults' perceptions of usefulness of personal health records. *Universal Access in the Information Society*, *12*(2), 191–204. https://doi.org/10.1007/s10209-012-0275-y

Prieto, G., & Leahy, D. (2012). Online social networks and older people. In K. Miesenberger, A. Karshmer, P. Penaz, & W. Zagler (Eds.), *Computers helping people with special needs* (Vol. 7382, pp. 666–672). Springer. https://doi.org/10.1007/978-3-642-31522-0_100

Prior, S., Arnott, J., & Dickinson, A. (2008). Interface metaphor design and instant messaging for older adults. In *CHI'08 extended abstracts on human factors in computing systems* (pp. 3747–3752). https://doi.org/10.1145/1358628.1358924

Procter, R., Greenhalgh, T., Wherton, J., Sugarhood, P., Rouncefield, M., & Hinder, S. (2014). The day-to-day co-production of ageing in place. *Computer Supported Cooperative Work (CSCW), 23*(3), 245–267. https://doi.org/10.1007/s10606-014-9202-5

Pruchno, R. (2019). Technology and aging: An evolving partnership. *The Gerontologist, 59*(1), 1–5. https://doi.org/10.1093/geront/gny153

Pullin, G., & Newell, A. (2007). Focussing on extra-ordinary users. In C. Stephanidis (Ed.), *Universal access in human computer interaction. Coping with diversity* (Vol. 4554, pp. 253–262). Springer. https://doi.org/10.1007/978-3-540-73279-2_29

Purao, S., Hao, H., & Meng, C. (2021). The use of smart home speakers by the elderly: Exploratory analyses and potential for big data. *Big Data Research, 25*, 100224. https://doi.org/10.1016/j.bdr.2021.100224

Purtilo-Nieminen, S., Vuojärvi, H., Rivinen, S., & Rasi, P. (2021). Student teachers' narratives on learning: A case study of a course on older people's media literacy education. *Teaching and Teacher Education, 106*, 103432. https://doi.org/10.1016/j.tate.2021.103432

Qian, Y., Schwartz, A. M., Zhang, Y., Jung, A., Wilds, G., Seitz, U., Kim, M., Kramer, A. F., & Chukoskie, L. (2024). Promoting cognitive health in older adults through an exercise game centered around foreign language learning. In *Extended abstracts of the CHI conference on human factors in computing systems* (pp. 1–7). https://doi.org/10.1145/3613905.3650959

Qian, Z., Fu, J., & Zhou, Y. (2024). Overcoming barriers, achieving goals: A case study of an older user's technology autonomy. In *Extended abstracts of the CHI conference on human factors in computing systems* (pp. 1–7). https://doi.org/10.1145/3613905.3637150

Qian, Z., Fu, J., & Zhou, Y. (2025). Exploring cultural and intergenerational dynamics in voice assistant design for Chinese older adults. *Proceedings of the ACM on Interactive, Mobile, Wearable and Ubiquitous Technologies, 9*(1), 1–18. https://doi.org/10.1145/3712275

Quan-Haase, A., & Elueze, I. (2018). Revisiting the privacy paradox: Concerns and protection strategies in the social media experiences of older adults. In *Proceedings of the 9th international conference on social media and society* (pp. 150–159). https://doi.org/10.1145/3217804.3217907

Quan-Haase, A., Martin, K., & Schreurs, K. (2016). Interviews with digital seniors: ICT use in the context of everyday life. *Information, Communication and Society, 19*(5), 691–707. https://doi.org/10.1080/1369118X.2016.1140217

Quan-Haase, A., Williams, C., Kicevski, M., Elueze, I., & Wellman, B. (2018). Dividing the grey divide: Deconstructing myths about older adults' online activities, skills, and attitudes. *American Behavioral Scientist, 62*(9), 1207–1228. https://doi.org/10.1177/0002764218777572

Rapoliene, G., Gedvilaite-Kordusiene, M., & Tretjakova, V. (2024). Barriers of information and communication technologies (ICT): Narratives of older users and their facilitators. *Romanian Journal of Communication and Public Relations, 25*(3), 7–23. https://doi.org/10.21018/rjcpr.2023.3.443

Rapp, A. (2023). Wearable technologies as extensions: A postphenomenological framework and its design implications. *Human–Computer Interaction, 38*(2), 79–117. https://doi.org/10.1080/07370024.2021.1927039

Rasi, P., Lindberg, J., & Airola, E. (2021). Older service users' experiences of learning to use eHealth applications in sparsely populated healthcare settings in Northern Sweden and Finland. *Educational Gerontology, 47*(1), 25–35. https://doi.org/10.1080/03601277.2020.1851861

Rasi, P., Vuojärvi, H., & Rivinen, S. (2021). Promoting media literacy among older people: A systematic review. *Adult Education Quarterly, 71*(1), 37–54. https://doi.org/10.1177/0741713620923755

Rasi, P., Vuojärvi, H., & Ruokamo, H. (2019). Media literacy for all ages. *Journal of Media Literacy Education, 11*(2), 1–19. https://doi.org/10.23860/JMLE-2019-11-2-1

Rassmus-Gröhn, K., & Magnusson, C. (2014). Finding the way home: Supporting wayfinding for older users with memory problems. In *Proceedings of the 8th Nordic conference on human–computer interaction: Fun, fast, foundational* (pp. 247–255). https://doi.org/10.1145/2639189.2639233

Read, J. C., & MacFarlane, S. (2006). Using the fun toolkit and other survey methods to gather opinions in child computer interaction. In *Proceedings of the 2006 conference on interaction design and children* (pp. 81–88). https://doi.org/10.1145/1139073.1139096

Reddy, G. R., Blackler, A., Popovic, V., Thompson, M. H., & Mahar, D. (2020). The effects of redundancy in user-interface design on older users. *International Journal of Human–Computer Studies, 137*, 102385. https://doi.org/10.1016/j.ijhcs.2019.102385

Renaud, K., & Van Biljon, J. (2010). Worth-centred mobile phone design for older users. *Universal Access in the Information Society, 9*(4), 387–403. https://doi.org/10.1007/s10209-009-0177-9

Renaud, K., & van Biljon, J. (n.d.). *Predicting technology acceptance and adoption by the elderly: A qualitative study.*

Rice, M., & Carmichael, A. (2013). Factors facilitating or impeding older adults' creative contributions in the collaborative design of a novel DTV-based application.

Universal Access in the Information Society, *12*(1), 5–19. https://doi.org/10.1007/s10 209-011-0262-8

Rice, M., Koh, R. Y. I., & Ng, J. (2016). Investigating gesture-based avatar game representations in teenagers, younger and older adults. *Entertainment Computing*, *12*, 40–50. https://doi.org/10.1016/j.entcom.2015.10.004

Richards, O. K. (2017). Exploring the empowerment of older adult creative groups using maker technology. In *Proceedings of the 2017 CHI conference extended abstracts on human factors in computing systems* (pp. 166–171). https://doi.org/10.1145/3027063. 3048425

Rienzo, A., & Cubillos, C. (2020). Playability and player experience in digital games for elderly: A systematic literature review. *Sensors*, *20*(14), 3958. https://doi.org/10. 3390/s20143958

Rivero Jiménez, B., Conde-Caballero, D., & Juárez, L. M. (2021). Loneliness among the elderly in rural contexts: A mixed-method study protocol. *International Journal of Qualitative Methods*, *20*, 1609406921996861. https://doi.org/10.1177/160940692199 6861

Rodrigues, N., & Pereira, A. (2021). User-centered rating of well-being in older adults. *IEEE Access*, *9*, 86824–86842. https://doi.org/10.1109/ACCESS.2021.3088748

Rodrigues, R., Huber, M., & Lamura, G. (n.d.). *Facts and figures on healthy ageing and long-term care.*

Roque, N. A., & Boot, W. R. (2018). A new tool for assessing mobile device proficiency in older adults: The mobile device proficiency questionnaire. *Journal of Applied Gerontology*, *37*(2), 131–156. https://doi.org/10.1177/0733464816642582

Rosales, A., & Fernández-Ardèvol, M. (n.d.). *Smartphones, apps and older people's interests: From a generational perspective.*

Rosales, A., Fernández-Ardèvol, M., Comunello, F., Mulargia, S., & Ferran-Ferrer, N. (2017). Older people and smartwatches, initial experiences. *El Profesional de La Información*, *26*(3), 457. https://doi.org/10.3145/epi.2017.may.12

Rosell, J., Sepúlveda-Caro, S., & Bustamante, F. (2024). Educational gerontechnology: Toward a comprehensive model for the education of digital technologies for older adults. In Q. Gao & J. Zhou (Eds.), *Human aspects of IT for the aged population* (Vol. 14725, pp. 275–292). Springer. https://doi.org/10.1007/978-3-031-61543-6_20

Ross, F., Donovan, S., Brearley, S., Victor, C., Cottee, M., Crowther, P., & Clark, E. (2005). Involving older people in research: Methodological issues. *Health and Social Care in the Community*, *13*(3), 268–275. https://doi.org/10.1111/j.1365-2524.2005.005 60.x

Rossetti, A., Cadwell, P., & O'Brien, S. (2020). "The terms and conditions came back to bite": Plain language and online financial content for older adults. In C. Stephanidis, M. Antona, Q. Gao, & J. Zhou (Eds.), *HCI international 2020: Late breaking papers: Universal access and inclusive design* (Vol. 12426, pp. 699–711). Springer. https://doi. org/10.1007/978-3-030-60149-2_53

Rossetti, A., Cadwell, P., & O'Brien, S. (n.d.). *Accessibility of online financial texts for ageing communities: An exploratory study.*

Russell, C. (1999). Interviewing vulnerable old people: Ethical and methodological implications of imagining our subjects. *Journal of Aging Studies, 13*(4), 403–417. https://doi.org/10.1016/S0890-4065(99)00018-3

Šabanović, S. (2010). Robots in society, society in robots: Mutual shaping of society and technology as a framework for social robot design. *International Journal of Social Robotics, 2*(4), 439–450. https://doi.org/10.1007/s12369-010-0066-7

Sachdeva, R. (n.d.). *Dynamics of the senior consumer market: A review and research agenda.*

Saldaño, V., Martin, A., Gaetán, G., & Vilte, D. (2013). *Web accessibility for older users: A southern argentinean view.*

Saldaño, V., Martin, A., Gaetán, G., & Vilte, D. (2014). *Focusing on older web users: An experience in patagonia argentina.*

Sames, K. M., & Hutson, J. A. (n.d.). *Adoption of smart home technologies: A qualitative exploration of older adults' and their family members' experiences.*

Samuel, G., Correa, M., Todd, C., & Tinker, A. (2023). What contribution do robots make to the care of older people? A scoping review. *Gerontechnology, 22*(2), 1–9. https://doi.org/10.4017/gt.2023.22.2.sam.08

Sánchez-Criado, T., López, D., Roberts, C., & Domènech, M. (2014). Installing telecare, installing users: Felicity conditions for the instauration of *Usership. Science, Technology, and Human Values, 39*(5), 694–719. https://doi.org/10.1177/016224391351 7011

Santos Silva, R., Mol, A. M., & Ishitani, L. (2019). Virtual reality for older users: A systematic literature review. *International Journal of Virtual Reality, 19*(1), 11–25. https://doi.org/10.20870/IJVR.2019.19.1.2908

Sarcar, S., Munteanu, C., Jokinen, J., Charness, N., Dunlop, M., Ren, X., & Waycott, J. (2020). Designing interactions for the ageing populations: Addressing global challenges. In *Extended abstracts of the 2020 CHI conference on human factors in computing systems* (pp. 1–5). https://doi.org/10.1145/3334480.3375164

Sardar, Z. (2010). The Namesake: Futures; futures studies; futurology; futuristic; foresight—What's in a name? *Futures, 42*(3), 177–184. https://doi.org/10.1016/j.futures. 2009.11.001

Sauvé, L., & Kaufman, D. (2020). User-centered design: An effective approach for creating online educational games for seniors. In H. C. Lane, S. Zvacek, & J. Uhomoibhi (Eds.), *Computer supported education* (Vol. 1220, pp. 262–284). Springer. https://doi. org/10.1007/978-3-030-58459-7_13

Sawa, Y., Keckeis, J., & Seaborn, K. (2023). Right for the job or opposites attract? Exploring cross-generational user experiences with "younger" and "older" voice assistants. *Designing Interactive Systems Conference, 124,* 160–163. https://doi.org/10.1145/ 3563703.3596642

Sawchuk, K., & Crow, B. (2010). Talking 'costs': Seniors, cell phones and the personal and political economy of telecommunications in Canada. *Telecommunications Journal of Australia, 60*(4), 55.1–55.11. https://doi.org/10.2104/tja10055

Sayago, S. (n.d.). *Kinect gestural UI: first impressions.*

Schehl, B., Leukel, J., & Sugumaran, V. (2019). Understanding differentiated internet use in older adults: A study of informational, social, and instrumental online activities. *Computers in Human Behavior, 97*, 222–230. https://doi.org/10.1016/j.chb.2019.03.031

Schiavo, G., Mich, O., Ferron, M., & Mana, N. (2022). Trade-offs in the design of multimodal interaction for older adults. *Behaviour and Information Technology, 41*(5), 1035–1051. https://doi.org/10.1080/0144929X.2020.1851768

Schlomann, A., Wahl, H.-W., Zentel, P., Heyl, V., Knapp, L., Opfermann, C., Krämer, T., & Rietz, C. (2021). Potential and pitfalls of digital voice assistants in older adults with and without intellectual disabilities: Relevance of participatory design elements and ecologically valid field studies. *Frontiers in Psychology, 12*, 684012. https://doi.org/10.3389/fpsyg.2021.684012

Sears, A., & Hanson, V. (2011). *Representing users in accessibility research.*

Seifert, A., & Schlomann, A. (2021). The use of virtual and augmented reality by older adults: Potentials and challenges. *Frontiers in Virtual Reality, 2*, 639718. https://doi.org/10.3389/frvir.2021.639718

Sengpiel, M. (2017). Teach or design? How older adults' use of ticket vending machines could be more effective. *ACM Transactions on Accessible Computing, 9*(1), 1–27. https://doi.org/10.1145/2935619

Sharit, J. (2020). The 'new' older worker. *Gerontechnology, 19*(2), 102–114. https://doi.org/10.4017/gt.2020.19.2.003.00

Shishehgar, M., Kerr, D., & Blake, J. (2018). A systematic review of research into how robotic technology can help older people. *Smart Health, 7–8*, 1–18. https://doi.org/10.1016/j.smhl.2018.03.002

Shiwani, T., Relton, S., Evans, R., Kale, A., Heaven, A., Clegg, A., Ageing data research collaborative (geridata) AI group.

Abuzour, A., Alderman, J., Anand, A., Bhanu, C., Bunn, J., Collins, J., Cutillo, L., Hall, M., Keevil, V., Mitchell, L., Ogliari, G., Penfold, G., et al. (2023). New horizons in artificial intelligence in the healthcare of older people. *Age and Ageing, 52*(12), afad219. https://doi.org/10.1093/ageing/afad219

Shneiderman, B. (2003). *Leonardo's laptop: Human needs and the new computing technologies* (1. MIT Press paperback ed.). MIT Press.

Short, A., Su, N. M., Hu, R., Choe, E. K., Kacorri, H., Danilovich, M., Conroy, D. E., Jette, S., Barnett, B., & Lazar, A. (2025). Tracking and its potential for older adults with memory concerns. In *Proceedings of the 2025 CHI conference on human factors in computing systems* (pp. 1–15). https://doi.org/10.1145/3706598.3714093

Siek, K. A., Rogers, Y., & Connelly, K. H. (2005). Fat finger worries: How older and younger users physically interact with PDAs. In M. F. Costabile & F. Paternò (Eds.),

Human–computer interaction: INTERACT 2005 (Vol. 3585, pp. 267–280). Springer. https://doi.org/10.1007/11555261_24

Silva, T., Caravau, H., & Carvalho, D. (2018). Comparative usability study of an iTV interface for seniors. In *Proceedings of the 8th international conference on software development and technologies for enhancing accessibility and fighting info-exclusion* (pp. 310–316). https://doi.org/10.1145/3218585.3218675

Silver, C. B. (2003). Gendered identities in old age: Toward (de)gendering? *Journal of Aging Studies, 17*(4), 379–397. https://doi.org/10.1016/S0890-4065(03)00059-8

Siriaraya, P., & Siang Ang, C. (2012). Age differences in the perception of social presence in the use of 3D virtual world for social interaction. *Interacting with Computers, 24*(4), 280–291. https://doi.org/10.1016/j.intcom.2012.03.003

Slegers, K., Van Boxtel, M. P. J., & Jolles, J. (2012). Computer use in older adults: Determinants and the relationship with cognitive change over a 6 year episode. *Computers in Human Behavior, 28*(1), 1–10. https://doi.org/10.1016/j.chb.2011.08.003

Sloan, D., Atkinson, M. T., Machin, C., & Li, Y. (2010). The potential of adaptive interfaces as an accessibility aid for older web users. In *Proceedings of the 2010 international cross disciplinary conference on web accessibility (W4A)* (pp. 1–10). https://doi.org/10.1145/1805986.1806033

Sonderegger, A., Schmutz, S., & Sauer, J. (2016). The influence of age in usability testing. *Applied Ergonomics, 52*, 291–300. https://doi.org/10.1016/j.apergo.2015.06.012

Sorgalla, J., Schabsky, P., Sachweh, S., Grates, M., & Heite, E. (2017). Improving representativeness in participatory design processes with elderly. In *Proceedings of the 2017 CHI conference extended abstracts on human factors in computing systems* (pp. 2107–2114). https://doi.org/10.1145/3027063.3053076

Soubutts, E., Singh, A., Ashcroft, A., Knowles, B., McDowell, J., Tsouvalis, J., Fledderjohann, J., Swarbrick, C., Harper, R., & Rogers, Y. (2025). Hidden opportunities for elder living: Understanding shared technology troubles and benefits for older adults in the UK cost of living crisis. In *Proceedings of the 2025 CHI conference on human factors in computing systems* (pp. 1–17). https://doi.org/10.1145/3706598.3714012

Staub, B., Doignon-Camus, N., Després, O., & Bonnefond, A. (2013). Sustained attention in the elderly: What do we know and what does it tell us about cognitive aging? *Ageing Research Reviews, 12*(2), 459–468. https://doi.org/10.1016/j.arr.2012.12.001

Stegner, L., Senft, E., & Mutlu, B. (2023). Situated participatory design: a method for in situ design of robotic interaction with older adults. In *Proceedings of the 2023 CHI conference on human factors in computing systems* (pp. 1–15). https://doi.org/10.1145/3544548.3580893

Stigall, B., & Caine, K. (2020). A systematic review of human factors literature about voice user interfaces and older adults. In *Proceedings of the human factors and ergonomics society annual meeting* (Vol. 64, pp. 13–17). https://doi.org/10.1177/1071181320641004

Stößel, C., & Blessing, L. (2010). Mobile device interaction gestures for older users. In *Proceedings of the 6th Nordic conference on human–computer interaction: extending boundaries* (pp. 793–796). https://doi.org/10.1145/1868914.1869031

Stößel, C., Wandke, H., & Blessing, L. (2010). Gestural interfaces for elderly users: Help or hindrance? In S. Kopp, & I. Wachsmuth (Eds.), *Gesture in embodied communication and human–computer interaction* (Vol. 5934, pp. 269–280). Springer. https://doi.org/10.1007/978-3-642-12553-9_24

Subasi, O., & Reithner, E. (2012). Needs and motivations of senior travelers for AAL. In *Proceedings of the 5th international conference on PErvasive technologies related to assistive environments* (pp. 1–4). https://doi.org/10.1145/2413097.2413111

Subasi, Ö., Fitzpatrick, G., Malmborg, L., & Östlund, B. (2013). Design culture for ageing well: Designing for 'situated elderliness'. In A. Holzinger, M. Ziefle, M. Hitz, & M. Debevc (Eds.), *Human factors in computing and informatics* (Vol. 7946, pp. 581–584). Springer. https://doi.org/10.1007/978-3-642-39062-3_36

Subasi, Ö., Malmborg, L., Fitzpatrick, G., & Östlund, B. (2014). Reframing design culture and aging. *Interactions, 21*(2), 70–73. https://doi.org/10.1145/2574561

Sum, S., Mathews, M. R., Pourghasem, M., & Hughes, I. (2008). Internet technology and social capital: How the internet affects seniors' social capital and wellbeing. *Journal of Computer-Mediated Communication, 14*(1), 202–220. https://doi.org/10.1111/j.1083-6101.2008.01437.x

Sustar, H. (2008). *Facilitating and measuring older people's creative engagement in a user centred design process.* People and Computers XXII Culture, Creativity, Interaction. https://doi.org/10.14236/ewic/HCI2008.94

Szymkowiak, A. (n.d.). *Issues surrounding the user-centred development of a new interactive memory aid.*

Takagi, H., Kosugi, A., Ishihara, T., & Fukuda, K. (2014). Remote IT education for senior citizens. In *Proceedings of the 11th web for all conference* (pp. 1–4). https://doi.org/10.1145/2596695.2596714

Talamo, A., Camilli, M., Di Lucchio, L., & Ventura, S. (2017). Information from the past: How elderly people orchestrate presences, memories and technologies at home. *Universal Access in the Information Society, 16*(3), 739–753. https://doi.org/10.1007/s10209-016-0508-6

Tetley, J., Grant, G., & Davies, S. (2009). Using narratives to understand older people's decision-making processes. *Qualitative Health Research, 19*(9), 1273–1283. https://doi.org/10.1177/1049732309344175

Thuesen, J., Feiring, M., Doh, D., & Westendorp, R. G. J. (2023). Reablement in need of theories of ageing: Would theories of successful ageing do? *Ageing and Society, 43*(7), 1489–1501. https://doi.org/10.1017/S0144686X21001203

Tinker, A. (2023). Four decades: A longstanding interest in technology and older people. *Gerontechnology, 22*(2), 1–3. https://doi.org/10.4017/gt.2023.22.2.tin.08

Tomás, L., & Ravazzini, L. (2022). Inclusiveness plus mixed methods: An innovative research design on transnational practices of older adults. *The Gerontologist, 62*(6), 816–822. https://doi.org/10.1093/geront/gnab128

Tortora, C., Di Crosta, A., La Malva, P., Prete, G., Ceccato, I., Mammarella, N., Di Domenico, A., & Palumbo, R. (2024). Virtual reality and cognitive rehabilitation for older adults with mild cognitive impairment: A systematic review. *Ageing Research Reviews, 93*, 102146. https://doi.org/10.1016/j.arr.2023.102146

Trewin, S. (with Association for Computing Machinery & ACM Special Interest Group on Accessible Computing). (2009). *Proceedings of the 11th international ACM SIGACCESS conference on computers and accessibility.* ACM. https://doi.org/10.1145/163 9642

Trewin, S., Richards, J. T., Hanson, V. L., Sloan, D., John, B. E., Swart, C., & Thomas, J. C. (2012). Understanding the role of age and fluid intelligence in information search. In *Proceedings of the 14th international ACM SIGACCESS conference on computers and accessibility* (pp. 119–126). https://doi.org/10.1145/2384916.2384938

Trucil, D. E., Lundebjerg, N. E., & Busso, D. S. (2021). When it comes to older adults, language matters and is changing: American geriatrics society update on reframing aging style changes. *Journal of the American Geriatrics Society, 69*(1), 265–267. https://doi.org/10.1111/jgs.16848

Tsai, H. S., Shillair, R., Cotten, S. R., Winstead, V., & Yost, E. (2015). Getting grandma online: Are tablets the answer for increasing digital inclusion for older adults in the U.S.? *Educational Gerontology, 41*(10), 695–709. https://doi.org/10.1080/03601277.2015.1048165

Tsai, W., Rogers, W. A., & Lee, C. (n.d.). *Older adults' motivations, patterns, and improvised strategies of using product manuals.*

Tse, M. M. Y., Choi, K. C. Y., & Leung, R. S. W. (2008). E-health for older people: The use of technology in health promotion. *CyberPsychology and Behavior, 11*(4), 475–479. https://doi.org/10.1089/cpb.2007.0151

Tuisku, O., Pekkarinen, S., Hennala, L., & Melkas, H. (2019). "Robots do not replace a nurse with a beating heart": The publicity around a robotic innovation in elderly care. *Information Technology and People, 32*(1), 47–67. https://doi.org/10.1108/ITP-06-2018-0277

Twigg, J., & Martin, W. (2015). The challenge of cultural gerontology. *The Gerontologist, 55*(3), 353–359. https://doi.org/10.1093/geront/gnu061

Ullal, A., Tauseef, M., Watkins, A., Juckett, L., Maxwell, C. A., Tate, J., Mion, L., & Sarkar, N. (2024). An iterative participatory design approach to develop collaborative augmented reality activities for older adults in long-term care facilities. In *Proceedings of the CHI conference on human factors in computing systems* (pp. 1–21). https://doi.org/10.1145/3613904.3642595

University of Wollongong, AU, Hasan, H., Linger, H., & Monash University, AU. (2018). Older women online: Engaged, active and independent. In *Australasian conference on information systems 2018*. University of Technology, Sydney. https://doi.org/10.5130/acis2018.cw

United Nations. (2020). *World population ageing 2020 Highlights: Living arrangements of older persons*. United Nations.

Upadhyay, P., Heung, S., Azenkot, S., & Brewer, R. N. (2023). Studying exploration and long-term use of voice assistants by older adults. In *Proceedings of the 2023 CHI conference on human factors in computing systems* (pp. 1–11). https://doi.org/10.1145/3544548.3580925

Vaisutis, K., Brereton, M., Robertson, T., Vetere, F., Durick, J., Nansen, B., & Buys, L. (2014). Invisible connections: Investigating older people's emotions and social relations around objects. In *Proceedings of the SIGCHI conference on human factors in computing systems* (pp. 1937–1940). https://doi.org/10.1145/2556288.2557314

Van Dijk, J. A. G. M. (2006). Digital divide research, achievements and shortcomings. *Poetics, 34*(4–5), 221–235. https://doi.org/10.1016/j.poetic.2006.05.004

Van Turnhout, K., Bennis, A., Craenmehr, S., Holwerda, R., Jacobs, M., Niels, R., Zaad, L., Hoppenbrouwers, S., Lenior, D., & Bakker, R. (2014). Design patterns for mixed-method research in HCI. *Proceedings of the 8th Nordic conference on human–computer interaction: Fun, fast, foundational* (pp. 361–370). https://doi.org/10.1145/2639189.2639220

Vassallo, M. (2024). The need for evidence-based mobile health technology. *Age and Ageing, 53*(3), afae034. https://doi.org/10.1093/ageing/afae034

Verbeek, P.-P. (2005). *What things do. Philosophical reflections on technology, agency, and design*. The Pennsylvania State University Press.

Verdezoto, N., & Grönvall, E. (n.d.). *Designing a tablet touch-screen interface for older adults*.

Vichitvanichphong, S., Talaei-Khoei, A., Kerr, D., & Ghapanchi, A. H. (2014). Assistive technologies for aged care: Supportive or empowering? *Australasian Journal of Information Systems, 18*(3), 880. https://doi.org/10.3127/ajis.v18i3.880

Vincent, J. A. (2006). Ageing contested: Anti-ageing science and the cultural construction of old age. *Sociology, 40*(4), 681–698. https://doi.org/10.1177/0038038506065154

Vošner, H. B., Bobek, S., Kokol, P., & Krečič, M. J. (2016). Attitudes of active older internet users towards online social networking. *Computers in Human Behavior, 55*, 230–241. https://doi.org/10.1016/j.chb.2015.09.014

Vroman, K. G., Arthanat, S., & Lysack, C. (2015). "Who over 65 is online?" Older adults' dispositions toward information communication technology. *Computers in Human Behavior, 43*, 156–166. https://doi.org/10.1016/j.chb.2014.10.018

Vuojärvi, H., Purtilo-Nieminen, S., Rasi, P., & Rivinen, S. (2021). Conceptions of adult education teachers-in-training regarding the media literacy education of older people: A phenomenographic study to inform a course design. *Journal of Media Literacy Education, 13*(3), 1–18. https://doi.org/10.23860/JMLE-2021-13-3-1

Wada, M., Hurd Clarke, L., & Mortenson, W. B. (2019). 'I am busy independent woman who has sense of humor, caring about others': Older adults' self-representations in online dating profiles. *Ageing and Society, 39*(5), 951–976. https://doi.org/10.1017/S0144686X17001325

Wagner, N., Hassanein, K., & Head, M. (2010). Computer use by older adults: A multi-disciplinary review. *Computers in Human Behavior, 26*(5), 870–882. https://doi.org/10.1016/j.chb.2010.03.029

Wakkary, R., Oogjes, D., Lin, H. W. J., & Hauser, S. (2018). Philosophers living with the tilting bowl. In *Proceedings of the 2018 CHI conference on human factors in computing systems* (pp. 1–12). https://doi.org/10.1145/3173574.3173668

Walker, A. (2007). Why involve older people in research? *Age and Ageing, 36*(5), 481–483. https://doi.org/10.1093/ageing/afm100

Wang, C., Yan, J., Huang, L., & Cao, N. (2024). Helping middle-aged and elderly short-video creators attract followers: A mixed-methods study on Douyin users. *Information Technology and People, 37*(3), 1305–1333. https://doi.org/10.1108/ITP-03-2022-0203

Wang, G., Chang, F., Gu, Z., Kasraian, D., & Van Wesemael, P. J. V. (2024). Co-designing community-level integral interventions for active ageing: A systematic review from the lens of community-based participatory research. *BMC Public Health, 24*(1), 649. https://doi.org/10.1186/s12889-024-18195-5

Wang, X., Knearem, T., & Carroll, J. M. (2019). Never stop creating: A preliminary inquiry in older adults' everyday innovations. In *Proceedings of the 13th EAI international conference on pervasive computing technologies for healthcare* (pp. 111–118). https://doi.org/10.1145/3329189.3329192

Wang, Y.-L. (2024). Technophilia or technophobia: The unified model of the paradox of Taiwanese older adults' digital learning. *Universal Access in the Information Society.* https://doi.org/10.1007/s10209-024-01184-1

Wang, Y., Li, M., Kim, Y.-H., Lee, B., Danilovich, M., Lazar, A., Conroy, D. E., Kacorri, H., & Choe, E. K. (2024). Redefining activity tracking through older adults' reflections on meaningful activities. In *Proceedings of the CHI conference on human factors in computing systems* (pp. 1–15). https://doi.org/10.1145/3613904.3642170

Wardt, V. V. D., Bandelow, S., & Hogervorst, E. (2013). The relationship between cognitive abilities, well-being and use of new technologies in older people. *Gerontechnology, 10*(4), 187–207. https://doi.org/10.4017/gt.2012.10.4.001.00

Warnock, D., McGee-Lennon, M., & Brewster, S. (2013). Multiple notification modalities and older users. In *Proceedings of the SIGCHI conference on human factors in computing systems* (pp. 1091–1094). https://doi.org/10.1145/2470654.2466139

Waycott, J., Pedell, S., Vetere, F., Ozanne, E., Kulik, L., Gruner, A., & Downs, J. (2012). Actively engaging older adults in the development and evaluation of tablet technology. In *Proceedings of the 24th Australian computer–human interaction conference* (pp. 643–652). https://doi.org/10.1145/2414536.2414633

Waycott, J., Vetere, F., Pedell, S., Morgans, A., Ozanne, E., & Kulik, L. (2016). Not for me: Older adults choosing not to participate in a social isolation intervention. In *Proceedings of the 2016 CHI conference on human factors in computing systems* (pp. 745–757). https://doi.org/10.1145/2858036.2858458

Weaver, C. K., Zorn, T., & Richardson, M. (2010). GOODS NOT WANTED: Older people's narratives of computer use rejection. *Information, Communication and Society, 13*(5), 696–721. https://doi.org/10.1080/13691180903410535

Weilenmann, A. (2010). Learning to text: An interaction analytic study of how seniors learn to enter text on mobile phones. In *Proceedings of the SIGCHI conference on human factors in computing systems* (pp. 1135–1144). https://doi.org/10.1145/1753326.1753496

Weiss, R. S., & Bass, S. A. (Eds.). (2001). *Challenges of the third age: Meaning and purpose in later life.* Oxford University Press. https://doi.org/10.1093/oso/9780195133394.001.0001

Werner, J. M., Carlson, M., Jordan-Marsh, M., & Clark, F. (2011). Predictors of computer use in community-dwelling, ethnically diverse older adults. *Human Factors: The Journal of the Human Factors and Ergonomics Society, 53*(5), 431–447. https://doi.org/10.1177/0018720811420840

Wiberg, M., & Stolterman, E. (2014). What makes a prototype novel? A knowledge contribution concern for interaction design research. In *Proceedings of the 8th Nordic conference on human–computer interaction: Fun, fast, foundational* (pp. 531–540). https://doi.org/10.1145/2639189.2639487

Wiles, J. L., Leibing, A., Guberman, N., Reeve, J., & Allen, R. E. S. (2012). The meaning of 'aging in place' to older people. *The Gerontologist, 52*(3), 357–366. https://doi.org/10.1093/geront/gnr098

Wilińska, M. (2015). An older person and new media in public discourses: Impossible encounters? In J. Zhou, & G. Salvendy (Eds.), *Human aspects of IT for the aged population. Design for aging* (Vol. 9193, pp. 405–413). Springer. https://doi.org/10.1007/978-3-319-20892-3_40

Williams, T. J., Jones, S. L., Luttcroth, C., Dekoninck, E., & Boyd, H. C. (2021). Augmented reality and older adults: A comparison of prompting types. In *Proceedings of the 2021 CHI conference on human factors in computing systems* (pp. 1–13). https://doi.org/10.1145/3411764.3445476

Wister, A., O'Dea, E., Fyffe, I., & D. Cosco, T. (2021). Technological interventions to reduce loneliness and social isolation among community-living older adults: A scoping review. *Gerontechnology, 20*(2), 1–16. https://doi.org/10.4017/gt.2021.20.2.30-471.11

Wöckl, B., Yildizoglu, U., Buber, I., Aparicio Diaz, B., Kruijff, E., & Tscheligi, M. (2012). Basic senior personas: A representative design tool covering the spectrum of European older adults. In *Proceedings of the 14th international ACM SIGACCESS conference on computers and accessibility* (pp. 25–32). https://doi.org/10.1145/2384916. 2384922

Wright, P. (2016). Helping older adults conquer digital tablets. *Gerontechnology, 14*(2), 78–88. https://doi.org/10.4017/gt.2016.14.2.005.00

Wu, A. Y., & Munteanu, C. (2018). Understanding older users' acceptance of wearable interfaces for sensor-based fall risk assessment. In *Proceedings of the 2018 CHI conference on human factors in computing systems* (pp. 1–13). https://doi.org/10.1145/ 3173574.3173693

Xie, B. (2011). Effects of an eHealth literacy intervention for older adults. *Journal of Medical Internet Research, 13*(4), e90. https://doi.org/10.2196/jmir.1880

Xie, B., Druin, A., Fails, J., Massey, S., Golub, E., Franckel, S., & Schneider, K. (2012). Connecting generations: Developing co-design methods for older adults and children. *Behaviour and Information Technology, 31*(4), 413–423. https://doi.org/10. 1080/01449291003793793

Xie, B., Watkins, I., Golbeck, J., & Huang, M. (2012). Understanding and changing older adults' perceptions and learning of social media. *Educational Gerontology, 38*(4), 282–296. https://doi.org/10.1080/03601277.2010.544580

Xie, C., Xie, Y., Wang, Y., Zhou, P., Lu, L., Feng, Y., & Liang, C. (2024). Understanding older adults' continued-use intention of AI voice assistants. *Universal Access in the Information.*

Xing, Y., Kelly, R. M., Rogerson, M. J., Waycott, J., & Aslam, K. (2024). Designing for inclusive experiences: Investigating opportunities for supporting older adults in community-based social programs. In *Proceedings of the CHI conference on human factors in computing systems* (pp. 1–20). https://doi.org/10.1145/3613904.3641892

Xu, T. B., Mostafavi, A., Kim, B. C., Lee, A. A., Boot, W., Czaja, S., & Kalantari, S. (2023). Designing virtual environments for social engagement in older adults: A qualitative multi-site study. In *Proceedings of the 2023 CHI conference on human factors in computing systems* (pp. 1–15). https://doi.org/10.1145/3544548.3581262

Yu, J. E., Parde, N., & Chattopadhyay, D. (2023). "Where is history": Toward designing a voice assistant to help older adults locate interface features quickly. In *Proceedings of the 2023 CHI conference on human factors in computing systems* (pp. 1–19). https:// doi.org/10.1145/3544548.3581447

Yu, J., Wu, J., Liu, B., Zheng, K., & Ren, Z. (2024). Efficacy of virtual reality technology interventions for cognitive and mental outcomes in older people with cognitive disorders: An umbrella review comprising meta-analyses of randomized controlled trials. *Ageing Research Reviews, 94*, 102179. https://doi.org/10.1016/j.arr.2023.102179

Yu, R. P., Ellison, N. B., McCammon, R. J., & Langa, K. M. (2016). Mapping the two levels of digital divide: Internet access and social network site adoption among older adults in the USA. *Information, Communication and Society, 19*(10), 1445–1464. https://doi.org/10.1080/1369118X.2015.1109695

Yuan, C. W., Hanrahan, B. V., Rosson, M. B., & Carroll, J. M. (2018). Coming of old age: Understanding older adults' engagement and needs in coproduction activities for healthy ageing. *Behaviour and Information Technology, 37*(3), 232–246. https://doi. org/10.1080/0144929X.2018.1432686

Zafrani, O., Nimrod, G., Krakovski, M., Kumar, S., Bar-Haim, S., & Edan, Y. (2024). Assimilation of socially assistive robots by older adults: An interplay of uses, constraints and outcomes. *Frontiers in Robotics and AI, 11*, 1337380. https://doi.org/10. 3389/frobt.2024.1337380

Zhang, S., & Boot, W. R. (2023). Predicting older adults' continued computer use after initial adoption. *Innovation in Aging, 7*(4), igad029. https://doi.org/10.1093/geroni/iga d029

Zhao, J. C., Davis, R. C., Foong, P. S., & Zhao, S. (2015). CoFaçade: A customizable assistive approach for elders and their helpers. In *Proceedings of the 33rd annual ACM conference on human factors in computing systems* (pp. 1583–1592). https://doi.org/10. 1145/2702123.2702588

Zhao, J., Chen, D., Barbareschi, G., & Sato, C. (2025). The role of ICT tools through a community program co-creation in a Japanese aging community. In *Proceedings of the 2025 CHI conference on human factors in computing systems* (pp. 1–10). https://doi. org/10.1145/3706598.3714158

Zhao, W., Kelly, R. M., Rogerson, M. J., & Waycott, J. (2023). Older adults using technology for meaningful activities during COVID-19: An analysis through the lens of self-determination theory. In *Proceedings of the 2023 CHI conference on human factors in computing systems* (pp. 1–17). https://doi.org/10.1145/3544548.3580839

Zhou, J., Salvendy, G., Boot, W. R., Charness, N., Czaja, S., Gao, Q., Holzinger, A., Ntoa, S., Rau, P.-L. P., Rogers, W. A., Stephanidis, C., Wahl, H.-W., & Ziefle, M. (2025). Grand challenges of smart technology for older adults. *International Journal of Human–Computer Interaction, 318*, 1–43. https://doi.org/10.1080/10447318.2025.245 7003

Zou, J., Liu, Z., & Zhao, C. (2024). Co-design for active aging: An approach to stimulating creativity of the young elderly in urban China. *The Design Journal, 27*(1), 89–110. https://doi.org/10.1080/14606925.2023.2275843

Zouba, N., Bremond, F., Thonnat, M., Anfosso, A., Pascual, É., Malléa, P., Mailland, V., & Guerin, O. (2009). A computer system to monitor older adults at home: Preliminary results. *Gerontechnology, 8*(3), 129–139. https://doi.org/10.4017/gt.2009.08. 03.011.00

Zuckerman, O., Walker, D., Grishko, A., Moran, T., Levy, C., Lisak, B., Wald, I. Y., & Erel, H. (2020). Companionship is not a function: The effect of a novel robotic object on healthy older adults' feelings of 'being-seen'. In *Proceedings of the 2020 CHI conference on human factors in computing systems* (pp. 1–14). https://doi.org/10.1145/3313831.3376411

GPSR Compliance
The European Union's (EU) General Product Safety Regulation (GPSR) is a set
of rules that requires consumer products to be safe and our obligations to
ensure this.

If you have any concerns about our products, you can contact us on

ProductSafety@springernature.com

In case Publisher is established outside the EU, the EU authorized
representative is:

Springer Nature Customer Service Center GmbH
Europaplatz 3
69115 Heidelberg, Germany

www.ingramcontent.com/pod-product-compliance
Ingram Content Group UK Ltd.
Pitfield, Milton Keynes, MK11 3LW, UK
UKHW050752080726
473054UK00005B/267